THE REBOUND EFFECT IN HOME HEATING

This is a definitive guide to the rebound effect in home heating – the increase in energy service use after a technological intervention aimed at reducing consumption. It sets out what the effect is, how it plays out in the home heating sector, what this implies for energy saving initiatives in this sector, and how it relates to rebound effects in other sectors. The book outlines how the concept of the rebound effect has been developed and the scope of research on it, both generally and particularly in the home heating sector. Within the context of energy and CO_2 emissions policy, it summarises the empirical evidence, exploring its causes and the attempts that are being made to mitigate it. Various definitions of the rebound effect are considered, in particular the idea of the effect as an energy efficiency 'elasticity'. The book shows how this definition can be rigorously applied to thermal retrofits, and to national consumption data, to give logically consistent rebound effect results that can be coherently compared with those of other sectors, and allow policymakers to have more confidence in the predictions about potential energy savings.

Ray Galvin has an interdisciplinary background, including engineering, social science and policy studies. He works with the Engineering and Architecture departments at the University of Cambridge, UK, and the Business and Economics Faculty at RWTH-Aachen University, Germany. His main research interest in recent years has been energy efficiency upgrades of existing homes, focusing mostly on Germany and the UK, but also on Europe more widely. He has published extensively on economic, social, technical and policy issues with regard to domestic energy consumption. His empirical and theoretical work includes research on the rebound effect in both buildings and transport.

◈BRI **Book Series**

New interdisciplinary and transdisciplinary approaches need a forum for information and discussion.

This book series shares similar aims and scope to the journal *Building Research & Information*, but allows for a deeper discussion, together with more practical material.

SCOPE: This book series explores the linkages between the built, natural, social and economic environments, with an emphasis on the interactions between theory, policy and practice. Emphasis is on the performance, impacts, assessment, contributions, improvement and value of buildings, building stocks and related systems: ecologies, resources (water, energy, air, materials, building stocks, etc.), sustainable development (social, economic, environmental and natural capitals) and climate change (mitigation and adaptation).

If you wish to contribute to the series then contact the series editor Richard Lorch at richard@rlorch.net with a short note about your ideas.

Forthcoming BRI series books:

The Rebound Effect in Home Heating
A guide for policymakers and practitioners
Ray Galvin

Towards a Recovery of Natural Environments in Architecture
C. Alan Short

Professionalism for the Built Environment
Simon Foxell

THE REBOUND EFFECT IN HOME HEATING

A GUIDE FOR POLICYMAKERS AND PRACTITIONERS

RAY GALVIN

First published 2016
by Routledge
2 Park Square, Milton Park, Abingdon, Oxon OX14 4RN

and by Routledge
711 Third Avenue, New York, NY 10017

Routledge is an imprint of the Taylor & Francis Group, an informa business

British Library Cataloguing in Publication Data
A catalogue record for this book is available from the British Library

Library of Congress Cataloging in Publication Data
Galvin, Ray.
The rebound effect in home heating : a guide for policymakers & practitioners / Ray Galvin.
pages cm. – (Routledge building research and information)
Includes bibliographical references and index.
1. Energy consumption. 2. Fuel trade. 3. Heating and ventilation industry–Government policy. 4. Energy policy. I. Title.
HD9502.A2G35 2015
333.79'63–dc23
2014049458

ISBN: [978-1-138-78834-3] (hbk)
ISBN: [978-1-138-78835-0] (pbk)
ISBN: [978-1-315-69694-2] (ebk)

Typeset in Univers
by Saxon Graphics Ltd, Derby

Printed and bound in the United States of America
by Edwards Brothers Malloy on sustainably sourced paper

CONTENTS

FIGURES AND TABLES

Figures

PREFACE

The rebound effect is an important subject because it causes energy savings and CO_2 emission reductions to be significantly less than expected. For the past 40 years governments have promoted energy efficiency improvements in home heating, electrical appliances, transport, industry and many other sectors of the economy. Initially this was to save energy, and more recently climate protection has also come to be a strong motivating factor. Energy efficiency has risen enormously in some sectors and significantly in others. Nevertheless, energy consumption and CO_2 emissions have not reduced as much as would be expected from these improvements. This is mostly due to the rebound effect. This book sets out to explain what the rebound effect is and how policymakers and other stakeholders can come to terms with it more effectively.

The rebound effect is a poorly understood concept. There are many different ways of measuring shortfalls in energy savings and giving numbers to the 'rebound effect'. Different approaches in research can produce widely different rebound effect percentages for the same situation, depending on what is being compared with what, and whether the definition of the rebound effect includes all the causes of energy savings shortfalls or only some. Values for the rebound effect can range from 5 per cent to 120 per cent for a sector such as private transport or domestic heating, simply because definitions of the rebound effect are different or not well thought through. This book offers a descriptive approach which clarifies definitions, and uses the two most robust and commonly used definitions of the rebound effect to map out a picture of energy savings short-falls in domestic heating: their magnitude, their causes, and how they may be better managed and mitigated. It also offers preliminary observations about rebound effects in commercial and other non-domestic buildings.

This book is for people who have a stake in energy saving in these buildings. This includes policymakers at national and local government level, the building and refurbishment industry, homeowners, housing providers, investors and building occupants, including tenants. The book is also intended for students and others in the academic arena who are learning about the rebound effect or undertaking research in this area.

To meet the needs and interests of this diverse range of readers, the book offers both a point-by-point, descriptive approach throughout, and a separate mathematical section in the Appendix. Some of the concepts pertaining to the rebound effect are challenging to grasp, and these are explained step by step in the narrative. Some of these concepts go hand in hand with mathematical proofs, but these are confined to the Appendix. The resulting formulas are used in the main text, as this is ineviin a study of a topic for which clear results and numbers are needed. However, readers who have little or no interest in formulas and equations, and are willing to take these calculations on trust, can skip over the formulas without losing the thread of the argument or the significance of the numbers.

The rebound effect is very important for energy planning and CO_2 emissions targets. Homeowners need to know how much their heating bill is likely to be reduced if they upgrade their homes according to various energy efficiency standards on offer. Housing providers and social agencies need to know what levels of investment are needed in the housing stock to eliminate fuel poverty. Architects and building engineers need to know how much energy is actually likely to be saved from retrofit features designed to produce given energy efficiency improvements. Policymakers need to know how great the average energy efficiency increase will need to be throughout the housing stock, in order to reduce energy consumption and CO_2 emissions by the large percentages being recommended by climate scientists. None of these can be known unless the likely rebound effects are understood, clearly defined and robustly calculated. The results of such calculations will seldom be absolute certainties, but if the rebound effect is properly understood they will be far better estimates than mere guesses or wishful thinking.

The book begins in Chapter 1 with an explanation of what the rebound effect is, including a brief history of its development as a concept, how it emerges in domestic heating, and a point-by-point explanation of the most prominent rebound effect definition, which is widely used by economists.

Chapter 2 discusses the causes of rebound effects in domestic heating in more detail, describing how occupant behaviour and technical difficulties combine to produce shortfalls in energy savings that are often both large and unexpected.

In the next two chapters, the logic and methodology of identifying and evaluating rebound effects are outlined. This begins in Chapter 3 with a description of a related concept known as the 'prebound' effect. This concept, only recently identified and defined, explains what is usually the largest portion of the shortfall in energy savings in domestic heating upgrades. It also provides a link between the two most commonly used, but widely different, definitions of the rebound effect, which are called in this book the 'elasticity rebound effect' and the 'energy performance gap'.

Chapter 4 sets out four different empirical methods for calculating rebound effects in domestic heating: one from case studies of individual dwellings and the others from large datasets of energy consumption and efficiency in buildings. The elasticity rebound effect proves to be consistent and robust for all these situations, while the energy performance gap also provides important insights.

Chapter 5 considers rebound effects in passive houses and low energy buildings. Rebound effects turn out to be high among these buildings, and the choice of rebound effect definition is very important in dealing with unique aspects of this range of energy consumption.

Chapter 6 explores the issue of fuel poverty in relation to the rebound effect. There are consistently high rebound effects when upgrading homes to mitigate fuel poverty. This is often seen as a problem by researchers and policymakers, but it is argued here that when definitions of the rebound effect are used carefully, much of this concern turns out to be unfounded.

Chapter 7 extends the scope of the discussion to explore rebound effects in upgrades of non-residential buildings. There are no existing studies on this topic, and those which touch on it are technical reports rather than peer-reviewed research papers. This chapter surveys related studies to date, offers some preliminary estimates of rebound effects, and discusses whether cross-learning might be possible between energy saving approaches in residential and non-residential buildings.

Chapter 8 gathers together the most important findings of the book. It draws conclusions, offers key insights and makes recommendations for policymakers, practitioners, other stakeholders and academics engaged in the field of rebound effect studies on domestic buildings.

The Appendix gives the formal mathematical derivations and proofs of the concepts and formulas used throughout the book. This is essential reference material for readers who wish to verify the credibility of the logic and concepts employed in the book. The relevant topics are numbered A.1, A.2, etc., and the equations are labelled (A1), (A2), etc. These are drawn upon or referred to as cross-references throughout the book for quick reference to relevant proofs and derivations. The Appendix can also stand separately as a short reference or primer for students and professionals undertaking their own studies of rebound effects.

ABBREVIATIONS AND SYMBOLS MOST FREQUENTLY USED IN THIS BOOK

C	calculated energy consumption, given in kWh/m²a
E	actual energy consumption, given in kWh/m²a
EPG	energy performance gap
ESD	energy savings deficit
EnEV	the German thermal building regulations (*Energieeinsparverordnung*)
kWh	kilowatt-hours
kWh/m²a	kilowatt-hours per square metre of living area per year
ln (*number*)	the natural logarithm of (the *number* in the brackets)
R	elasticity rebound effect (energy efficiency elasticity of energy services)
P	prebound effect
SAP	Standard Assessment Procedure (a UK indicator of the level of energy efficiency of domestic heating systems
TWh	Terawatt-hours (1 TWh = 1,000,000,000 kWh)
ε	energy efficiency (the Greek letter *epsilon*)
η	elasticity (the Greek letter *eta*)

Further and more specialised symbols are defined and explained in the Appendix

1

THE REBOUND EFFECT AND DOMESTIC HEATING

1. Setting the scene

In the six months from October 1973 to March 1974 the price of oil quadrupled, from US$3 to US$12 per barrel. Political issues, together with declining oil production in the United States, led to concern about high and unstable oil prices continuing into the future (Bardi, 2009). As one of a number of measures to enhance the security of energy supply, OECD countries initiated a regulatory process for higher energy efficiency standards in buildings, household appliances and other sectors of the economy.

The economist Daniel Khazzoom investigated the effects of these energy efficiency improvements on the consumption of energy in household appliances. In theory, an increase in energy efficiency should cause a corresponding decrease in energy consumption: doubling the energy efficiency should halve the energy consumption. In a paper that has become one of the most frequently cited in the field of energy studies, Khazzoom (1980) reported that in fact, the decrease in energy consumption was consistently less than predicted. He suggested increases in energy efficiency were not being fully utilised to produce savings in energy consumption. Instead, only a portion of these, if any, was being utilised for this purpose. The remaining portion was being syphoned off to increase the consumption of energy 'services'.

Energy 'services' are the benefits people get from consuming energy: hours of watching television; indoor warmth on a cold winter's day; kilometres (km) travelled in a car; products rolling off an assembly line. Khazzoom suggested people were increasing their consumption of energy services, as it was now cheaper to do so, since their appliances were now more energy efficient. Higher energy efficiency meant the price per unit of energy services had fallen. Consumers were able to take more energy services than previously, yet still pay less overall.

Khazzoom warned that energy efficiency improvements could therefore 'backfire' and lead to an increase in energy consumption, particularly in industry. Higher energy efficiency reduced the price of factory goods, which increased sales, leading to higher profits, providing an incentive for manufacturers to install more plant and produce even more goods.

The same phenomenon was observed by economist Len Brookes, working independently from Khazzoom (Brookes, 1990). Both called the phenomenon 'backfire' and both were concerned that energy efficiency improvements, which were aimed to reduce energy consumption, could have the opposite effect.

In later papers, Khazzoom (1987; 1989) noted that this phenomenon had been observed in the nineteenth century by the English philosopher and economist William Stanley Jevons. Jevons (1865) drew attention to the paradox that the more efficiently Britain consumed coal, the more coal it consumed – an observation that became known as the 'Jevons paradox' (see discussion in Alcott, 2005). In nineteenth century Britain, increasing the efficiency of steam engines and their associated machinery brought down energy costs per unit of production, making goods cheaper, leading to more sales, which increased manufacturers' returns, enabling them to invest in more machinery and afford yet more efficient machines, thus consuming more coal to produce more goods. Increasing energy efficiency therefore led to more energy consumption.

Two prominent researchers of the 1980s disagreed strongly with Khazzoom's and Brookes' 'backfire' theory. Amory Lovins (1988) maintained that these counterproductive effects of energy efficiency improvements would be 'insignificantly small', though it is not clear what empirical evidence was offered for this claim. Grubb (1990) repeated this assertion, but also pointed to an important distinction between different types of energy efficiency improvements. There is, he suggested, a natural, continuous improvement in energy efficiency which is driven by the scarcity of resources and the desire for greater well-being. This will certainly lead to higher energy consumption, as its goal is economic growth. However, energy efficiency improvements that are specifically designed to save energy will not necessarily lead to significantly higher demand, he maintained, as they are not part of this growth cycle. They are designed to enable consumers to enjoy the same level of benefits as before, but for a lower price, consuming less energy.

Both Lovins and Grubb agreed that this will lead to some shortfall in the expected energy savings, but maintained this would be insignificantly small and would not lead to backfire.

It was in the midst of this debate that the terms 'rebound' and 'rebound effect' emerged. There is no certainty as to when precisely these terms began to be used. Lovins spoke of 'rebound' in this context in a brief paper in 1988 (Lovins, 1988), and the term 'rebound effect' was used in a conference discussion as early as 1983. Berry and Hirst (1983) reported that a conference delegate had spoken of a 'rebound effect' caused by householders who 'raise their thermostat settings after completing conservation retrofits' on their homes (p. 78). This is a good example of one common cause of the rebound effect in domestic heating.

By the early 1990s the study of this phenomenon had entered mainstream energy and economics research, initially under the rubric of the 'Khazzoom–Brookes postulate', a label coined by Saunders (1992). During the 1990s the use of the terms 'rebound effect' and 'backfire' became standardised in academic literature. 'Backfire' came to be used for cases where an energy efficiency improvement led to an increase in energy consumption. The more subtle situation was called the 'rebound effect' – where there was a reduction in energy consumption, but not as much as would have been expected, given the size of the energy efficiency improvement.

A further point introduced by Khazzoom (1980, 1989) was that the rebound effect can reverberate throughout an entire economy. An increase of energy savings and energy services in one sphere, such as manufacturing, can lead to cheaper products and more spare cash, which can be spent on other energy consuming activities. This 'economy-wide' or 'macroeconomic' rebound effect has been extensively discussed and investigated in more recent years, since computing power became available to track rebound effects in detail through-out the major sectors of an economy (e.g. Barker *et al.*, 2007), or indeed the world economy (Barker *et al.*, 2009).

2. Domestic heating, energy efficiency, energy services

2.1 Energy efficiency

One of the key terms in understanding the rebound effect is 'energy efficiency'. Energy efficiency is a measure of how effectively the consumption of fuel is turned into the benefits that human beings want. Some devices can be given a precise, absolute energy efficiency figure. For example, a modern condensing boiler typically achieves 90 per cent efficiency. This means that 90 per cent of the energy consumed in the fuel driving the boiler is successfully converted into heat to warm the building. Older, conventional boilers typically have effi-ciencies of around 70–80 per cent, and the efficiency of wood stoves has risen from 30 per cent to up to 90 per cent.

Other devices and systems cannot be given an absolute efficiency figure, but instead are given figures relative to each other. A house or apartment block is an example of this. A house acts as a composite system, with energy being brought in and consumed to heat it, and the heat then being lost to the outside world by conducting and radiating out through walls, floor, roof and windows, or drifting out through cracks and openings. Efficiency for a dwelling is therefore a more complex concept than for just a boiler or wood stove. The better a dwelling retains its heat, the more efficient it is. For example, a dwelling which requires 100 kilowatt-hours of energy consumption per square metre of floor area per year (kWh/m²a) to keep a steady, indoor temperature of 20°C is twice as efficient as a dwelling that requires 200kWh/m²a to keep the same indoor temperature.

Throughout this book, energy efficiency is almost always treated as a rela-tive metric, rather than an absolute metric. Since the book will always be dealing with *changes* in energy efficiency relative to what they were before the change, this makes good mathematical sense and simplifies things greatly (see Appendix, Section A.1).

The efficiency of home heating has been steadily increasing over the last 40 years, particularly since the first oil shock in 1973–74, and even more sharply since concern about climate change became widespread late last century. A home built to the minimum standard of the building code in most western European countries in 2014 is about four times as efficient as one built to the minimum standard in the mid-1970s, when countries first included energy effi-ciency requirements in their building codes (Galvin, 2012). This means it is designed to produce the same level of thermal comfort for one-quarter of the energy consumption. In Germany for example, the building regulations in 1977 set the maximum permissible energy consumption for domestic heating at an average of 270kWh/m²a. This was reduced progressively over the following decades, until the most recent tightening of the regulations in 2009 brought it to 70kWh/m²a. The energy efficiency requirement for German homes today is therefore 3.86 times as high as it was (or 2.86 times higher than it was) in 1976.

2.2 Energy services

A further concept that is essential for understanding the rebound effect is that of 'energy services'. Energy services are the benefits that people derive from consuming energy. In the home these include a large number of benefits which can be identified under the broad headings of thermal comfort, electronic entertainment, cooked food, cleaned clothes and dishes, and the operation of devices that enable people to work from home. These are not hard and fast divisions. Cooking is often part of entertainment, while computers can be used for both work and entertainment. The benefits relating to domestic heating come under the broad heading of thermal comfort. This includes air and radiant heating, evenness of warmth, air humidity level, air purity, heating up and cooling down time, and level of breeze or draught. Cooling is also an aspect of thermal comfort, but is a subject in itself and is not covered in this book.

Energy services (such as thermal comfort) are the reason energy is actually consumed by people. When planning a new building or an energy efficiency upgrade on an existing home, engineers need to know what level of energy services they are aiming to achieve. Governments therefore set certain standards of thermal comfort, which new buildings have to attain. Many of these are based on or draw upon the standards published by ASHRAE (American Society of Heating, Refrigerating and Air-Conditioning Engineers), which attempt to define the level of indoor temperature, light, humidity, etc., which is comfortable and healthy for the average person (see discussion in Humphreys and Hancock, 2007). The German government has its own set of standards for thermal comfort, published by the German Institute for Standards (DIN – *Deutsche Institut für Normung*). These stipulate an indoor temperature of at least 19°C in all rooms all year round, with three full exchanges of indoor air per hour. A building which achieves these standards may be said to be providing 100 per cent energy services. All new builds have to be designed to achieve these conditions (100 per cent energy services) in Germany's climatic conditions, without consuming more than the maximum level of energy consumption set down in the regulations for a building of that size and shape.

Energy services cannot be measured directly, as energy consumption can. If a building is not providing this level of energy services – say the indoor temperature is only 18°C and there is only one full exchange of air per day – it is very difficult to say what percentage of the full requirement of energy services it is providing.

A simple rule of thumb is used to address this problem. First, an energy performance rating (EPR) is calculated. This is the level of heating energy consumption that is required for this building to provide 100 per cent energy services. For new buildings, this will already have been calculated at the design phase, so as to conform to the regulations outlined above. Second, the building's actual heating energy consumption, averaged over the last two years and adjusted for any unusual climate conditions over those years, is noted. The average level of energy services is then defined as the actual consumption divided by the EPR. If, for example, the EPR was 100 kilowatt-hours per square metre of floor area per year (kWh/m^{2}a) and the actual consumption was 80kWh/m^{2}a, the level of energy services is taken to be 0.8, or 80 per cent.

This is not a perfect measure because it does not account for possible variations in types of thermal comfort. Nor does it account for the possibility that many people adapt to different temperatures, ventilation rates, etc., (see discussion in Chappells and Shove, 2007). However, as Galvin (2014a) explains in more detail, it provides a very useful rule of thumb for putting a number to the level of thermal comfort, and therefore energy services, being provided by a heating system and enjoyed by building occupants.

2.3 Energy efficiency and energy services

To understand the concept of the rebound effect, both energy and energy services have to be taken into account. These are two very different parameters. Energy services are the benefits people enjoy. Energy is the commodity that is consumed to provide these benefits.

Things can become confusing because the word 'consumption' applies to both. People 'consume' energy services: they 'use up' heat because it comes out of the radiator and escapes through the walls and windows, and they 'use up' fresh air by making it stuffy so that it has to be replenished. They also consume energy: they 'use up' the energy in the gas, oil or electricity that powers their heating systems. It is the interplay between these two commodities – energy services and energy – plus the parameter of energy efficiency, that leads to the rebound effect.

It was mentioned in 2.1 that new homes in Germany today are just under four times as energy efficient as homes built in 1977. However, today's new homes in Germany do not consume a mere one-quarter of their 1977 counterparts' consumption. Recent research indicates they consume, on average, about half the 1977 level (Schröder *et al.*, 2011; Sunikka-Blank and Galvin, 2012; Walberg *et al.*, 2011). The main reason for this discrepancy is the rebound effect. A sizeable portion of the energy *efficiency* increase has been used to increase the level of energy *services* (higher indoor temperatures, longer heating periods, more rooms heated, more generous ventilation), so that only the remaining portion of the energy efficiency upgrade has led to reductions in *energy* consumption. This is conceptually illustrated in Figure 1.1.

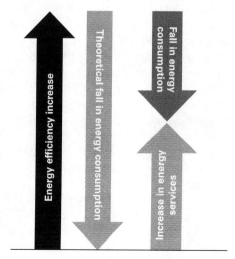

Figure 1.1
Schematic of the rebound effect

Starting from the left in Figure 1.1, the first arrow represents the increase in energy efficiency that has been achieved through upgrading a dwelling. The second arrow from the left represents the reduction in energy consumption that would have been expected with this magnitude of energy efficiency increase. These two arrows are shown the same size, though the relationship between them is actually reciprocal: a doubling of energy efficiency leads (theoretically) to a halving of energy consumption.

The next set of arrows show what the increase in energy efficiency *actually* leads to. A portion of it, represented by the top arrow, leads to a *reduction* in *energy* consumption, while the remaining portion, represented by the bottom arrow, leads to an *increase* in *energy services* consumption: higher temperatures, more generous ventilation, etc. It is these two things together that comprise the rebound effect: a *reduction* (in energy consumption) and an *increase* (in energy services consumption).

A further factor is that there are often technical failures with energy efficiency upgrades. Insulation might not cover walls consistently, boilers may fail to achieve their specified efficiencies, the heat gain from the sun might have been wrongly calculated. There are often also failures due to poor interfaces between occupants and heating controls (i.e. poorly explained or understood procedures for operation; poor feedback on when the system is working as intended; etc.). These factors can cause further shortfalls in energy savings. They can be very large, and are included in discussion of the rebound effect throughout this book. The important point here is that the rebound effect occurs because the increase in energy efficiency does not all lead to energy consumption reductions. Part of it is 'lost'. Some of these losses go to increasing energy services consumption, part to technical failings and part to failures in human interaction with the new (heating or control) technology.

The rebound effect typically occurs when dwellings undergo upgrades to increase their energy efficiency. In literature on the subject, such energy efficiency improvements are called 'thermal retrofits'; 'energy efficiency upgrades'; 'energy efficiency refurbishment'; or various combinations of these terms. This book uses all these terms interchangeably.

In some cases of energy efficiency refurbishment, the rebound effect is so great that 'backfire' occurs (see Section 1): more energy is consumed after the refurbishment than prior to it. A typical example is where a household has an upstairs floor in their house that is so difficult and expensive to heat that they almost never use it, and the radiators are switched off for most of the time. After refurbishment the upstairs floor becomes easy and cheap to heat. The result is that a household may expand its daily activities and keep the upstairs rooms warm, with the result that the house consumes more heating energy than it did before. Actual examples of backfire in UK dwellings after refurbishment are noted in Hong *et al.* (2006), and discussed in Chapter 6.

There can also be instances of negative rebound effects, which are often called 'super-conservation'. Here, energy consumption after a thermal retrofit reduces by a greater amount than would be expected if all the energy efficiency increase were used to reduce energy consumption. Intuitively this seems highly improbable, but a typical example would be where a household goes heavily into debt to pay for a thermal retrofit, and afterwards decides to economise by cutting down drastically on home heating. Negative rebound effects can also happen when a household increases their environmental awareness at the time they retrofit their home, and thereafter opts for a more frugal heating regime. Over the long term, it appears that in some European countries, reductions in heating energy consumption are running ahead of what would be expected as a result of energy efficiency upgrades, producing negative rebound effects (see Chapter 4).

3. Defining the rebound effect

A special issue of the journal *Energy Policy* in 2000 (see the editorial: Schipper, 2000) brought together leading contributions on the theme of the rebound effect, with broad agreement as to its definition, which may be stated:

> *The rebound effect occurs where an energy **efficiency** increase leads to an increase in the consumption of energy **services**.*

Given this starting point, the rebound effect can be defined from three different angles: as a phenomenon that occurs in certain ways and for certain reasons; as a mathematical relationship between energy efficiency, energy consumption and energy services consumption; and as an interplay of causes and their effects. Each of these is important and casts its own light on the issues.

3.1 The rebound effect as a phenomenon

Most academic discussion of the rebound effect has been led by economists, and one of the clearest expositions of what the rebound effect is, from an economics point of view, was offered by Berkhout *et al.* (2000). Essentially, economists define rebound as a *price* effect. An increase in energy efficiency lowers the cost of energy services, so people consume more services. This happens on both the *microeconomic* and *macroeconomic* levels. The microeconomic level has to do with the particular devices or appliances that are made more energy efficient, and how their owners and users react to this increased efficiency. On this micro-level there is a *direct* rebound effect and an *indirect* rebound effect. Taking the example of thermally retrofitting a home, the direct rebound effect is where the householder keeps higher indoor temperatures, heats more rooms in the house for longer periods, and opens the windows for longer periods, because the cost per unit of indoor warmth is lower. Another example is a car owner upgrading to a more energy efficient model. The direct rebound occurs if he or she drives the car further because the cost per kilometre is lower.

The *indirect* rebound effect, however, occurs when savings made through an energy efficiency upgrade in one type of technology are used to consume more energy in another type of technology. After a thermal retrofit a household may use its heating bill savings to finance such items as increased computer use in the home, more air travel for holidays (greater distances and/or more frequent holidays), or purchases of novel electronic devices. Indirect rebound effects are particularly relevant to home heating because householders can make large savings through comprehensive thermal retrofits, while new possibilities of consumption are constantly on offer. An oft cited example is the household who can finally afford to fly from northern Europe to the Mediterranean for a sunny holiday because their energy efficiency improvements in their home saved them so much in heating bills. In Chitnis *et al.*'s (2013) words, they turn 'lights into flights'.

One of the perverse things about indirect rebound effects is that they also tend to occur when ecologically conscious householders reduce their heating bills by heating more frugally and strategically, even if these people do not retrofit their home. In an interesting study, Druckman *et al.* (2011) investigated how people's spending habits changed when they set out to reduce their CO_2 emissions by reducing indoor temperatures, not wasting food, and travelling less by car. Their average indirect rebound effect was 34 per cent, and one household's was as high as 515 per cent. The unit of measurement was kilograms of CO_2 emitted, rather than units of energy, but the principle is similar. Heating, eating and travel became cheaper, so a portion of the savings were redeployed into other CO_2-emitting practices.

It is possible, however, that some indirect rebound effects are not merely a 'price' effect. If environmentally conscious people feel they are doing the environment a good turn by reducing emissions in one sector, they may feel they have permission to be lax in other sectors. Further, the increased efficiency of devices such as computers is often also associated with technological innovations that lead people to have and use more electronic devices – not because they are energy efficient, but because they have become popular. The question of what is *causing* an increase in energy services consumption is very important, and is discussed in Section 3.3 below.

It can be very difficult to track and estimate indirect rebound effects. There are many options for people to spend the money saved through improving a home heating system. If it is left in the bank, it is available for others to borrow, to invest in other energy consuming activities. This book is primarily concerned with direct rebound effects in domestic heating, but will comment on indirect effects where these are relevant to its central theme.

3.2 The rebound effect as a mathematical relationship

3.2.1 THE MATHEMATICS EXPLAINED

From a mathematician's point of view the rebound effect, as it is defined by economists, is a very elegant and satisfying set of relationships. The detailed mathematics are set out in the Appendix of this book, while the concepts are explained in a descriptive and practical way throughout the main text. The themes and argument of the book can be understood without having to follow the mathematics in the Appendix, and for readers who prefer not to delve into the mathematics, the findings are presented in the main text.

The rebound effect is what economists call an *elasticity*. An elasticity is a special kind of relationship between two variables which are changing in relation to each other. It gives a measure of how much one variable changes when the other changes. This can provide a very useful basis for comparison of changes in different commodities or activities that are related in some significant way.

For example, the energy *efficiency* of the British housing stock is increasing every year, as old homes are made more energy efficient and new, highly energy efficient homes are built. At the same time, the average indoor temperature in British houses is rising every year, i.e. the level of energy *services* is increasing. Meanwhile the total energy *consumption* for heating British homes is reducing. These are three variables, all changing in relation to each other: energy efficiency, energy services and energy consumption. Any two of these variables can have their relationship to each other mapped over time as an elasticity, to give a clear, precise, simple measure of how they are changing in relation to each other. To begin with, the changes in energy efficiency and energy consumption are considered.

An elasticity is defined as:

The proportionate change in one variable, as a ratio of the proportionate change in another variable.

Therefore the 'energy efficiency elasticity of energy consumption' can be considered as follows. In the period 2001–2011 the average energy *efficiency* of homes in Britain *increased* by an average annual rate of 1.80 per cent. This represents a proportionate annual change of 1.0180: the energy efficiency in any given year was 1.0180 times the previous year's energy efficiency.

Over the same period the average energy *consumption* of a British home *reduced* by an average annual rate of 1.43 per cent (taking variations in the weather into account). This was a proportionate annual change of 0.9857: the average consumption in any given year was 0.9857 times that of the previous year (based on figures from the EU database Odyssee, 2013, and see further analysis in Chapter 4).

These two proportionate change figures can be brought together as an elasticity, to see at a glance how the energy consumption changed in comparison with the changes in energy efficiency, i.e. how *elastic* the energy consumption was in response to changes in efficiency. This is called the 'energy efficiency elasticity of energy consumption'. This figure is found simply by dividing the logarithm of the consumption figure by the logarithm of the efficiency figure (see Appendix, Section A.3):

$$\log 0.9857 \div \log 1.0180 = -0.8074$$

In prosaic terms, this means that for every 1 per cent increase in energy efficiency there was a 0.8074 per cent reduction in energy consumption (the reason logarithms are used in these elasticity calculations is explained in the Appendix).

This implies that there was a rebound effect. If there were no rebound effect, every 1 per cent increase in energy efficiency would lead to a 1 per cent reduction in energy consumption. The answer to the above sum would have been −1.000 instead of −0.8074 per cent.

This leads to a second combination of the proportionate change figures, the energy efficiency and the energy *services* consumption. This is called the 'energy efficiency elasticity of energy services consumption'. It is in fact the same thing as the rebound effect (Sorrell and Dimitropoulos, 2008). It is found by dividing the logarithm of the energy services figure by the logarithm of the energy efficiency figure.

The problem here, of course, is that there is no energy services figure. The annual proportionate change in energy services is not given in the Odyssee database. This is not surprising, since energy services values are notoriously difficult to obtain directly. Obtaining them would require monitoring indoor temperature, window opening times and air quality in every room of a large random sample of homes in Britain, at least hourly, every day for at least one year.[1]

However, it is still possible to calculate a rebound effect figure for this data, due to a simple mathematical fact. The rebound effect is the energy efficiency elasticity of energy consumption plus 1.

$$\text{rebound effect} = 1 + \eta_{\varepsilon E} \qquad (1.1)$$

The symbol η (the Greek letter *eta*) is used to mean 'elasticity'. The subscript ε (the Greek letter *epsilon*) stands for 'efficiency' and E stands for 'energy consumption'.

This very elegant relationship can be quite simply proven mathematically (see Appendix, Section A.2), and was made widely known by Sorrell and Dimitropoulos (2008) in their detailed exposition of the key mathematical features of the rebound effect.

Using this formula, the rebound effect in the UK housing stock is $-0.8074 + 1 = 0.1926$, or 19.26 per cent. Expressed more prosaically, this means that about 19 per cent of the gain in energy efficiency in the UK housing stock was taken back and used to increase energy services, so that only the remaining 81 per cent was utilised to reduce energy consumption.

Of course, the figures given above for the British housing stock are merely averages. There were large variations between households, and the annual rates of change were not smooth. The figures of 1.0180 and 0.9857 above were obtained, first, by averaging out the changes in all the houses over each particular year, then by smoothing out the year-by-year fluctuations using a curve-fitting technique known as 'exponential least squares regression', which is explained in the Appendix, Section A.7.

3.2.2 THE MATHEMATICS OF THE REBOUND EFFECT AS A POLICY TOOL

Although these averages mask large home-by-home variations and some annual fluctuations, they provide very important information for policymakers, and this is an area where the mathematics of the rebound effect become very useful. Many European countries, including the UK, have adopted the EU policy of reducing CO_2 emissions by 80 per cent by the year 2050. To achieve this in domestic heating, the UK would have to reduce emissions by 4 per cent per year for 40 years.[2] If emissions were proportional to energy consumption, it would look as though this could be achieved by increasing the housing stock's energy efficiency by about 4 per cent per year. But when the rebound effect is taken into account, a mathematical analysis shows that energy efficiency actually has to be increased by over 5 per cent per year to achieve the 80 per cent reduction goal (see the mathematical derivations in Galvin, 2014b).

In Chapter 4 rebound effects are calculated in the same way as above for all 28 EU countries plus Norway. Knowing the rebound effect for each country can give a quick, rule-of-thumb measure of how much needs to be done to reduce emissions to the target level. For some countries, such as Malta, rebound effects are extremely high, while for most western European countries they range from 10–50 per cent. Britain would need to more than treble its current rate of energy efficiency increase and maintain that higher rate for the next 35 years to meet its CO_2 obligations by 2050. This could be offset somewhat by replacing carbon-rich fuels such as coal and oil with less carbon intensive fuels such as natural gas, and even more effectively with renewables. But even with these advantages, policymakers need to take the rebound effect into account, as it reduces the savings from these improvements as much as from energy efficiency gains.

3.2.3 ALTERNATIVE DEFINITIONS OF THE REBOUND EFFECT

Most discussion of the rebound effect accepts the economists' definition of it as an elasticity. A weakness of this 'elasticity rebound effect', as it is called throughout this book, is that it only works where *changes* in both efficiency and consumption are known. The above example compares *year-by-year* changes in efficiency with year-by-year changes in consumption. Chapter 4 shows how *one-off* changes in efficiency and consumption can also be used to calculate the elasticity rebound effect in specific cases where individual dwellings are thermally retrofitted. Chapter 4 also shows how similar calculations can be made when there are *implied* changes in efficiency and consumption, which can be estimated if there are large enough datasets of efficiency and consumption in dwellings at any particular time.

But there are alternative definitions of the rebound effect, often used by engineers, where there is not enough information to estimate these changes. A classic situation is where a new home is built, and the occupants consume significantly more heating energy than was calculated in the design. In this situation there are no *changes* in efficiency or consumption, but only one value for each of these variables. In this situation engineers often use a version of the rebound effect which in this book is called the 'energy performance gap': the amount of over-consumption, divided by the expected, calculated consumption, expressed as a percentage. This comes with a completely different set of mathematics from the 'elasticity' rebound effect (see Appendix, Section A.8), and is in fact a different metric entirely. However, with careful thinking it can also be used to inform stakeholders how much *better* their buildings have to be in order to meet CO_2 emission goals. In Chapter 4 this definition of the rebound effect and its uses are explored, along with those of the elasticity rebound effect and one further definition, which is called in this book the 'energy savings deficit'.

3.3 Interplay between causes, effects and the rebound effect

There is much discussion as to what *causes* the rebound effect. Under the strictest version of the economist's 'elasticity' definition, rebound is a 'price' effect. Economists distinguish between *exogenous* fuel prices, which are the prices consumers pay the suppliers, and *endogenous* fuel prices, which are the prices they pay per unit of energy services. Because the endogenous price of fuel is lower after an energy efficiency upgrade, householders consume it more liberally. This understanding enables economists to estimate the size of the rebound effect indirectly, by seeing how households respond to changes in the *exogenous* price of fuel, i.e. the price they pay the supplier. One such example is Madlener and Hauertmann (2011), where panel data on household heating energy consumption and heating fuel price were used to estimate rebound effects in Germany. The results were lower than, but within the range of, estimates which used methods more directly related to energy efficiency and consumption, such as those in Galvin (2014a, 2014b, 2015).

However, close investigation of particular households indicates there is much more than a price effect at play when households increase their consumption of energy services after an energy efficiency upgrade. First, a dwelling *functions differently*, in its thermal aspects, after being heavily insulated. Where previously it might have been hard to get the indoor temperature to reach 17°C, now it often reaches 18°C or 19°C without the boiler running. The household is not making their home warmer in response to cheaper fuel; they are not *making* it warmer at all, it just *is* warmer. This is a purely *physical* effect, not a price effect.

Second, often an energy efficiency upgrade includes replacement of old heating controls with newer, more sophisticated thermostats and timers. Many householders find these very difficult to use, and often adjust them poorly, resulting in more energy being consumed than is necessary to get the indoor conditions (i.e. thermal comfort) they want. This may be called a *user interface* effect, again rather than a price effect.

Third, some new heating technologies work efficiently only when operated as they were intended to be. However, they may not fit well with householders' daily rhythms of heating and ventilation needs. One such example is underfloor heating, which can take up to 24 hours to heat up or cool down. This can be convenient for people who cannot go out or move about much in their home, but for physically active people it can be wasteful because their periods of being at home and out cannot be easily matched to the periods when the heating is on. This may be called a *socio-technical mismatch* effect, and again it is not a price effect.

Finally, there may be faults in the technology itself: cavity wall insulation that does not fill all the gaps; boilers poorly matched to the size of the dwelling; heat losses due to poor build quality, such as gaps in insulation, poor air-tightness in the building envelope. These may be called *technology failure* effects.

It seems, then, that there are many different reasons why energy savings do not match energy efficiency increases. Some writers seek to frame this problem by making a two-fold distinction, between 'rebound effects' and 'technical failures'. In this framework, price effects only are defined as rebound effects, and other causes of energy saving shortfall are grouped together as technical failures. In this author's experience such a framework is inadequate, as it only touches the two extreme ends of what is a very wide spectrum of different sorts of closely interwoven effects. In interviews with homeowners and technical assessments of retrofits conducted by this author, it has rarely been possible to draw clear lines between the different types of effects listed above. In some cases, high consuming households can be identified and offered help in sorting through the often tangled causes of their high consumption (Galvin, 2013). At the extreme technical end of the spectrum, it can be useful to identify specific flaws in retrofit technology, such as cavity walls that are not filled properly (Hong, *et al.*, 2006) or lofts that are not properly sealed and insulated (Galvin and Sunikka-Blank, 2014a). It can also be useful to identify mismatches between heating technology and human occupants, in the hope that retrofit design will become more user-friendly. But often there is such an interweaving of human and technical effects that it is impossible to separate them into distinct categories.

Confining the rebound effect to price effects implies a misunderstanding of the dynamics of how occupants interact with their dwellings. After a retrofit, a dwelling is a different environment from before, and the household-plus-dwelling make up a different entity. The notion of a 'socio-technical system' (MacKenzie and Wajcman, 1985) can provide a useful framework for understanding this. Socio-technical systems theory maintains that human beings and technology are so closely interwoven that they are best understood as a composite entity, rather than two separate domains. This approach has contributed much to our understanding of how and why so much energy is consumed in homes and how this can be mitigated (e.g. Fawcett *et al.*, 2013; Killip, 2013). In such a framework, a researcher would not begin by conceiving user effects and technological failure as separate domains, but rather by being attentive to how various mixes of human and technical interactions produce their own effects.

This approach leads to a single figure for the rebound effect rather than a figure disaggregated into human and technical causes. A single figure can also be a very useful way of indicating to policymakers and other stakeholders that such-and-such a percentage of energy efficiency gains tend to be lost when dwellings are thermally upgraded. This information in itself may not help solve the 'problem' of the rebound effect, but it will give a clearer indication as to what level of energy efficiency planners and policymakers have to aim for, to get the reductions in energy consumption and the increases in energy services that society decides it wants.

Notes

1   There are of course studies which do such monitoring on a smaller scale. However, the Odyssee data, referred to here, has the advantage that its energy consumption figures take all dwellings into account.

2   At first sight the figure of 4 per cent might look like an error, but these are *cumulative* percentages. Reducing emissions by 4 per cent per year means that each year's emissions are 0.96 times the previous year's. 0.96 *to the power of* 40 = 0.2, or 20 per cent. Reducing emissions to 20 per cent of their 2010 level by 2050 equates to an 80 per cent reduction in 40 years.

2

WHAT CAUSES THE REBOUND EFFECT IN HOME HEATING?

1. Introduction: What counts as a rebound effect?

Chapter 1 suggested that there is a variety of quite different actions, failures and phenomena that can lead households to consume more heating energy after an energy efficiency upgrade than engineers expect or calculate. This chapter summarises the results of field studies over the last 15 years which indicate the range and diversity of these effects. The author of this book was directly involved in a number of these studies, while others come from a broader range of researchers.

The two main definitions of the rebound effect were noted in Chapter 1: an *elasticity*, which compares the proportionate *change* in energy services consumption with the proportionate change in energy efficiency; and an *energy performance gap*, which compares *actual* consumption with *expected* or calculated consumption. It will become clear that some of the effects discussed in this chapter can only be analysed with one or other of these definitions. This chapter is concerned with the causes of rebound effects, and Chapter 4 will look more closely at the range of definitions of the rebound effect and their differences.

Disagreements exist about what can legitimately be called the 'rebound effect' and what should be called something else. Under the strictest economists' definition, only *price* effects are rebound effects, meaning over-consumption can be called the rebound effect only when it is caused by households consuming more because it is now cheaper to do so.

Other definitions are somewhat more liberal, saying that as long as the consumers (and not the technology) are causing the over-consumption, this is the rebound effect. This includes such things as households' inability to adjust their new heating controls; using rooms which were previously too cold to bother heating; or staying at home and heating for more hours because it is now so comfortable at home.

Some definitions go yet further and include technical failings among the causes of the rebound effect. Others take these out of the calculation of the rebound effect, because they need to be rectified differently – by improving technology, rather than helping householders to consume more efficiently.

All these effects are counted as rebound effects in this book, for two main reasons. First, empirical research shows that the division between consumer behaviour and technological failure is not clear-cut (Galvin, 2013; Hong *et al.*,

2006). Often, in fact, the upgrading of a dwelling's heating technology can lead to consumer behaviour problems, as the thermal properties of a dwelling can become misaligned with the needs of householders. The notion of a 'socio-technical system' was introduced in Chapter 1 as a useful framework for considering all these effects under one rubric.

Second, regardless of what is causing consumption to be greater than expected, policymakers, engineers, local authorities, homeowners and other stakeholders need comprehensive information as to what reductions in consumption can be expected in the real world when buildings or retrofits are designed to achieve certain levels of energy efficiency. While policy often does have different initiatives for different spheres, such as retrofit technology as distinct from consumer behaviour, overall policy commitments typically include broad-brush figures, such as an 80 per cent reduction in CO_2 emissions by 2050. Handing policymakers a single figure for the average rebound effect in domestic heating energy can simplify their task of reaching agreement on aims, amidst diverse political forums. For example, it helps communication when policymakers know that, to achieve 80 per cent reductions, they actually have to aim for something closer to 88 per cent reductions because of losses due to the rebound effect.

The causes of rebound effects are discussed in this chapter under six main headings: comfort-taking; unplanned lifestyle changes; failures at the human–technology interface; technology miscalculations; technology failures; and miscalculations in efficiency modelling. While these divisions are conceptually separable, it will be seen that in practice there are significant crossovers between them.

2. Comfort-taking

By far the most common result of energy efficiency upgrades in homes is an increase in indoor temperature. This may seem obvious and expected, but it is a major cause of the rebound effect. This is because the calculated heating energy consumption before and after a retrofit assumes that the indoor temperature remains the same after the retrofit (compared to before). This comes clearly into focus when a highly regimented, finely calculated retrofit framework in building regulations is examined, such as the one in Germany.

In Germany it is illegal to do any energy efficiency upgrades on a home unless they are designed to achieve the very high standards laid down in the 'Energy Saving Regulations', popularly called the 'EnEV' (*Energieeinsparverordnung;* see EnEV, 2009). Further, if homeowners are repairing 10per cent or more of any relevant feature of their dwelling, such as a wall or roof, they are legally obliged to retrofit it to these standards. This demand is justified on the basis that the standards required are 'economically viable' (*wirtschaftlich*), i.e. the money the energy efficiency measures cost will pay back, through fuel savings, over their technical lifetime. The calculations the German government commissioned, to prove that retrofitting to these standards always pays back, clearly state that this assumes there will be no increase in indoor temperature after a retrofit (Hauser *et al.*, 2012; Kah *et al.*, 2008; and see discussion in Galvin, 2012; 2014c; Galvin and Sunikka-Blank, 2013). This leads to the situation that any increase in indoor temperature after a retrofit represents a reduction in the gains made by increasing the energy efficiency. The same applies to increased ventilation or any other improvement in indoor thermal comfort. Increased thermal comfort *is* a part of the rebound effect. It is often called 'comfort-taking', a phrase that is frequently used synonymously with 'rebound effect'.

Many policymakers in Germany have failed to acknowledge and account for comfort-taking. This inevitably will skew the figures for the amount of energy saved. Comfort-taking also thwarts the plans of engineers to reduce energy consumption through improving energy efficiency.

In Britain, however, comfort-taking is often seen in a positive light, because it represents an improvement in indoor living conditions for people whose homes were too cold and therefore unhealthy. In short, it combats fuel poverty. In a major study on the rebound effect in home heating in Britain, Milne and Boardman (2000) brought together the results of monitored retrofits from the last two decades of the twentieth century. They found a consistent positive correlation between indoor temperature and thermal efficiency features such as cavity wall insulation, double-glazed windows and loft insulation. Further, they found that in most retrofit cases, a portion of the energy efficiency increase went to reduce energy consumption while the remaining portion was taken back as increased indoor temperature. The portions varied depending on how cold the house had been before the upgrade. On average, for a house with 16°C indoors before an upgrade, about 30 per cent of the energy efficiency improvement was taken back in increased indoor temperature, a considerable quantity of 'comfort-taking'. The lower the pre-upgrade temperature of a dwelling, the higher the proportion that is taken back.

This is framed positively throughout Milne and Boardman's study, as the elimination of fuel poverty. Increasing a dwelling's heating energy efficiency brings a double dividend: reduction in energy consumption and warmer, healthier indoor conditions. The impact of Boardman's wider work on fuel poverty (e.g. Boardman, 2010) is partly responsible for the fact that thermal retrofit policy in Britain consistently brings these two goals together (see further discussion in Chapter 6). One important consequence is that a high proportion of UK public subsidies for retrofits are directed at low income households living in thermally poor dwellings. This has a further benefit: the first steps in raising the energy efficiency of thermally poor houses are inexpensive and bring large efficiency gains, an issue explored in Jakob (2006) and Galvin (2010).

Nevertheless, comfort-taking does not only occur when the thermally poorest homes are retrofitted to higher thermal standards. It seems to occur with energy efficiency upgrades in almost every class of home, even those which are relatively efficient to start with (cf. Summerfield *et al.*, 2010). In almost every research interview this author has conducted with households who have thermally retrofitted – in Britain, Germany, New Zealand and Austria – householders have mentioned that they now enjoy a much warmer indoor environment than before. Many wear light clothing at home in mid-winter and keep the thermostat at well over 20°C. Some older people keep the whole house very warm as they do not move around much, and do not want to risk the chill of going into a cold room even if it is little used and they enter it only briefly, once or twice a day. In general it seems that wherever an energy efficiency upgrade occurs, a significant portion of the energy efficiency gain will be lost to comfort-taking.

3. Unplanned lifestyle changes

When a dwelling is thermally retrofitted its indoor character changes, even if there are no changes in layout or furnishings. Cold niches become warm; rooms that were impossible to heat become comfortable. If insulation is substantial, often the entire house has an ambient temperature that makes economical heating of any particular room a viable option. When more rooms are warm, householders move about differently indoors. They spend more time sitting in rooms and corners they previously avoided, and find new uses for rooms which had been unusable in winter. In some cases they go out less often, spend less time in bed and more at a computer desk, entertain guests more often, gravitate more toward home hobbies and less toward outdoor activities. These changes are not always planned or expected prior to an energy efficiency upgrade, but are often consequences of it.

In a set of interviews with households who had retrofitted old homes in Britain to significantly higher thermal standards, at least half were clearly occupying their dwelling more intensively than previously (Galvin and Sunikka-Blank, 2014a). One had installed an electric heater in a loft, as this had become useable as a workroom now that the ceiling was insulated. A large Victorian home with two occupants was now heated throughout in every room, and daily activities had expanded to fill most of the house. A room with a piano was used more frequently because the householder could play without his fingers 'freezing'. Another household had suffered such severe dampness problems that it was often unpleasant to be at home, but now with insulation, mechanical ventilation and a more effective heating system, more rooms were used for longer periods.

Not all increases in occupier intensity after a retrofit are unplanned. One of the households interviewed in the research project mentioned above planned an upgrade of the heating system knowing this would bring two unusable bedrooms up to the standard where she could rent them to lodgers. Another household redesigned an unused part of their house as a guest suite for hosting itinerant members of an international religious organisation. Another made a cold, unusable room warm so that it could be used as home office space.

Indoor lifestyle changes such as these, planned or unplanned, cause more energy services to be consumed after a retrofit than prior to it. Therefore they are regarded here as causes of the rebound effect.

A further effect has a subtle relation to the rebound effect: many retrofits accompany building extension work. Almost half the households in the interview set mentioned above included building extensions in their project. Five extended their kitchen. One added a bedroom. One enlarged two sides of the house, giving an extra bathroom, utility room and all-purpose family room. A number of recent studies indicate there is often no clear line between thermal retrofitting and doing more general home improvements (Aune, 2007; Risholt and Berker, 2013; Tweed, 2013; Wilson *et al.*, 2013). Often in research interviews it is found that people start with one of these in mind and move on to include both, and very often the floor area of the home is increased as part of the process. One household, for example, started planning a retrofit because the house was bitterly cold (it was on the north side of a semi-detached pair), and ended up with a much larger as well as more modern home (Sunikka-Blank and Galvin, 2014).

Where the energy services consumption in a home increases due to the floor area increasing, this is not strictly the rebound effect, as it is the result of increased building area rather than higher energy efficiency. However, for practical purposes it can sometimes be included in the rebound effect because it involves real energy consumption that would probably not have happened without the stimulation of energy efficiency improvements. If greater accuracy is needed for calculating rebound effects in particular cases, consumption can be calculated according to floor area (or volume).

4. Failures at the human–technology interface

Thermal technology for homes consists of two main elements: insulation and a heating system. Insulation is passive, in the sense that once it is installed the occupants cannot adjust or optimise it. The heating system, however, is active, responding to both its own temperature and time sensors and the interventions of the user. At this interface major problems often occur and large rebound effects can follow.

The most common problem is the user's inability to adjust thermostats and timers. In most modern British and Continental heating systems there is one central thermostat and timer, plus thermostatic controls on each radiator. Operation is intended to be simple, but in interviews with households in Britain and on the Continent a complex tangle of problems is found (cf. Stevenson and Rijal, 2010).

The main thermostat might not be in a room that is representative of household heating needs. A thermostat in a hallway with draughts will signal the boiler to keep running until the hallway reaches the target temperature, thus running the boiler too high. Conversely, a thermostat in a much-used dining room will reach its target and switch off the boiler before other rooms are warm. A household will therefore need to set the thermostat high to get the radiators on in the other rooms, but then have to cool the dining room by opening windows to let the excess heat out. These difficulties cause rebound effects in homes where previously there was no central heating system (compare Hong et al., 2006). They push heating consumption above what it would have been if the household was always able to get its desired temperatures in each part of the house.

The situation can become more contorted when highly sophisticated digital thermostat-timers are installed. In some interviews in Continental Europe, this author's research has found that even householders who are engineers cannot work out how to operate the controls effectively. This almost always leads to over-heating, as householders put controls on or near their maximum settings. Close examination of sensor data from an estate of 90 apartments in a technologically cutting-edge retrofit project in Germany indicated that many households were using their thermostats as on–off switches, setting them on maximum to warm up the apartment and minimum to cool it down (Galvin, 2013).

Problems may also be due to lack of adequate, easily understandable information from the supply side on how to use and optimise the controls.

Further, in most homes there is a simple, hand-operated tap on each radiator for local thermostatic control. But some new installations dispense with this and instead have an electronic timer-thermostat on the wall in each room.

Many occupants express frustration that this further complicates their heating control, as these, too, can be difficult to operate effectively. In a number of research interviews in Austria, occupants have indicated their frustration dramatically by making the hand-action one uses for adjusting a traditional radiator, while pointing out that they knew what they were doing with such devices and always got it right. This has the ironic effect that over-consumption, and therefore rebound effects, can be caused by technology specifically designed to reduce over-consumption.

A further development in retrofits, as also in new housing, is the use of underfloor heating rather than radiators. In theory underfloor heating has two great advantages over radiators. First, it provides warmth in the right places: on the floor, where cold-sensitive feet tread, and evenly throughout the room. Second, as underfloor heating covers the entire area of each room, it can operate at a much lower temperature than radiators need to, while providing the same amount of heating energy in a room. Instead of heating water to 55°C to enable radiators to provide an indoor air temperature of 22°C, the water in underfloor heating needs to be heated to only around 30°C. Even though the same amount of useful work is produced, the laws of thermodynamics dictate that less energy is consumed by heating a greater quantity of water to 30°C, than a smaller quantity to 55°C. For this reason many engineers, environmental lobbyists and policy actors see underfloor heating as the energy efficient norm for the future (Wiltshire, 2011; and see discussion in McCrae, 2008).

But underfloor heating also brings two serious problems. First, it has a large 'thermal inertia', meaning it takes a long time to heat up and cool down. Lag times range from four hours to over 24 hours. This is suitable for buildings such as hospitals and the homes of people who are chronically ill or infirm, where steady, evenly distributed warmth is needed constantly. But most households do not need their whole dwelling comfortably warm all the time. Research interviews show there are large variations in households' thermal needs with respect to time of day, day of week, and work versus recreational periods, not to mention rapidly shifting thermal needs from room to room. A musician, for example, needs a high room temperature while sitting still watching television, but a lower temperature while playing the violin. Research interviewees report it seldom takes more than 20 minutes for the room temperature to adjust, when using traditional radiators. With underfloor heating, however, occupants often need to open the windows to cool the room sufficiently for a change from sitting still to moderately vigorous activity. This wastes a portion of the energy which underfloor heating was meant to save, adding to the rebound effect.

To date there have been no systematic studies of quantified energy losses through this phenomenon in homes, but studies were conducted on UK schools with underfloor heating (DfES, 2004; Moncaster, 2012). It was found that underfloor heating 'responds too slowly to react to the fast changes of utilisation in a school and therefore requires a supplementary form of heating' (DfES, 2004: 21). Very large rebound effects were found in schools that had installed underfloor heating.

A further effect of underfloor heating is that households become reluctant to turn off their heating when they go out, and even when going away for the weekend or longer, due to the fear of being cold for many hours while it heats up after their return. This can cause heating systems to run for significantly longer than they did prior to a retrofit.

Some engineers suggest this should not be a problem, as underfloor heating is designed to be kept running constantly, and at such a low temperature that not much energy will be wasted if no-one is home to enjoy the warmth. This contradicts the laws of physics, however, as the heat loss of a building is directly proportional to the difference between indoor and outdoor temperatures. The entire building warms up when underfloor heating is left on, as it heats the indoor air, which heats the walls and ceilings. In western European climates a rule of thumb is that every 1°C increase in indoor temperature leads to a 5–10 per cent increase in heating consumption, and this applies regardless of how good the building's insulation is (Hens, 2012).

A second reason rebound effects occur with underfloor heating is that *average* energy consumption is not reduced through the effect of households who choose to heat more sparingly. Regardless of the type of heating, there is always a proportion of households who over-consume, either by keeping high indoor temperatures, heating for excessively long periods, over-ventilating, or various combinations of these (Gram-Hanssen, 2010). For homes without underfloor heating this is counterbalanced by those who heat more sparingly, so that the average is somewhere near the median, and not close to the excessive range.

For example, very accurate studies of comparative consumption are possible with sets of passive houses, as these are all designed to consume 15kWh/m^2a for space heating (Peper and Feist, 2008). Although average actual heating consumption in passive houses is well above this (Galvin, 2014d; Schneiders and Hermelink, 2006), in all the datasets of passive house heating consumption to hand, heating consumption follows a roughly normal distribution skewed, with low consumers offsetting high consumers about one-to-one. The distribution is skewed to the right as expected, since there is a limit to how low consumption can go, but not to how high.

However, the early results of research suggest that this offsetting is less likely with underfloor heating, since households who claim to have been formerly low consumers have no opportunity to fine-tune their temperature adjustment from hour to hour and day to day. This lifts average consumption, and would have the effect of skewing the distribution even more strongly to the right, putting average consumption well above median consumption. No systematic studies have been carried out to confirm or disprove this, but such studies would fill an important gap.

Hence the interfaces between technology and householders can cause rebound effects where the technology fails to fit with household skills, lifestyle, rhythms of coming and going, or natural variations in over- and under-consumption habits.

5. Technology miscalculations

The various components of an energy efficient retrofit or new dwelling must be correctly calculated to interface together in a way that is optimised for minimum energy consumption. For example, many retrofitted homes show rebound effects because the boiler is too large for their occupants' needs. Running a large boiler at under-capacity can consume more energy than running a small boiler at full capacity. It can be very difficult to anticipate these needs prior to a retrofit, and designers often do not want to install a boiler that is below capacity and cannot reach the indoor temperature occupants demand. It is therefore important to plan boiler capacity as precisely as possible for each individual situation.

More serious mismatches can occur with heat pumps, including both ground- and air-source. A heat pump is akin to a refrigerator operating in reverse. It takes heat from the (cold) outdoors and transfers it into the (warmer) indoors. A heat pump operating at its optimum capacity can typically shift about three times as much heat energy from outdoors to indoors, as it consumes in electrical energy – a 'coefficient of performance' (COP) of 3.0. Heat pumps are often praised as a carbon-saving option for the future. Bergman *et al.* (2009) claim a COP of 4.1 is typically possible for domestic space heating.

The downside of a heat pump, however, is that its electricity consumption requires about 2.7 times that amount of energy to produce it at the generator. Therefore it is crucial to match components so that the heat pump functions at its full COP.

The COP of a heat pump is very sensitive to its loading, i.e. the amount of heat energy it is called upon to produce. The COP falls sharply as a heat pump's loading rises above its design level. This brings the opposite problem incurred by boilers: an *over*-consuming household can cause inefficiencies in a heat pump, leading to rebound effects. A series of studies in the UK showed that heat pumps used for space heating in a range of types of building typically achieved COPs in the range 1.2–3.2, with an average well below the break-even point of 2.7 (Moncaster, 2012).

The problem is exacerbated when heat pumps are used for underfloor heating. This is largely because, for the reasons outlined above, underfloor heating often runs well above its design capacity. Mismatches also occur when heat pumps power other forms of heating where the occupants have difficulty controlling their thermostats and timers. Their extra demands for heating energy force the heat pump to work above its optimum capacity.

There can also be problems matching heat pumps to their energy source. This may be moderately cold water in horizontal 'slinky' pipes laid just below the ground surface, or warmer deep bore water. The COP falls as the water temperature falls. Temperatures near the ground surface can vary significantly, and can fall significantly as the heat pump extracts heat, via this water, from the soil. However, it should be noted that there are often problems with measuring the performance of ground source heat pumps (Stafford, 2011).

Some of the cascading rebound effects due to these factors were illustrated in a study of three blocks of 30 apartments each in Germany (Galvin, 2014a). The first block was retrofitted with standard, traditional radiators powered by district heating. It was designed to consume $50kWh/m^2a$, and after two years of operation, measurements showed its apartments consumed an average of $51kWh/m^2a$, an energy performance gap of 2 per cent. The second block used underfloor heating, powered by district heating. It was designed to consume $37kWh/m^2a$, but consumed $58kWh/m^2a$, an energy performance gap of 57 per cent. The third building used heat pumps as its heating energy source. This powered underfloor heating in one-third of its apartments, radiant ceiling heating in another third and air heating (though the ventilation system) in the final third. These apartments were designed to consume 23.6kWh/m²a, but two years of measurements showed they were consuming 88kWh/m²a, an energy performance gap of 273 per cent. It is statistically extremely unlikely that the differences were caused merely by variations in consumer behaviour between residents in the three buildings. In this retrofit project, the higher the level of ambition above a basic retrofit, the higher the rebound effect.

6. Technology failures

Insulating homes to high standards is a very exacting task. A layer of 16cm of polystyrene insulation attached to the exterior of a wall has a U-value of 0.22, so that it insulates as effectively as a dry concrete wall 1m thick. In mid-winter this relatively thin layer often has to maintain a temperature difference between the inside and outside wall surfaces of 20°C or more. If there are gaps in the insulation, heat will conduct easily through these sections and energy will be lost. Moisture is likely to condense on the resulting indoor cold spots, leading to an increased need for ventilation to keep indoor humidity low, causing more heat loss. These losses cause more energy to be consumed than was planned and calculated for a retrofit, adding to the rebound effect.

There is a severe shortage of the building skills needed to meet the exacting standards required to avoid this kind of technology failure in retrofits, not to mention new builds. In Germany, for example, where regulations demand very high thermal standards and significant progress has been made in retrofits, interviews with building professionals and building support agencies have indicated that many failures occur through shortage of skills (Galvin, 2011). On a theoretical level, building physics in Germany is highly advanced, through the work of leading institutions such as the engineering faculty at the Technical University of Munich (www.bgu.tum.de/), the Fraunhofer Institute for Building Physics (www.ibp.fraunhofer.de/) and the Building Institute at the Technical University of Darmstadt (www.massivbau.tu-darmstadt.de), where the passive house was largely developed. But there is a large gap between the precision of these technical solutions, and the level of skill and know-how of tradespeople. The popular press has recently brought to light some of the common failures in this technical side of retrofitting, which cause significant over-consumption (e.g. Hanf, 2013) and rebound effects.

These problems are also occurring in Britain, despite standards being less demanding. A recent research project in Cambridge, for example, found wall external insulation that did not reach right to the end of the wall, loft insulation with a large gap above the bathroom ceiling, gaps of over 1cm in the seal of a loft trapdoor and cavity wall insulation with a gap of 1m around the entire house (Galvin and Sunikka-Blank, 2014a). Similar failings are reported by Hong et al. (2006). Some homeowners reported that if they had not been on-site when certain aspects of their retrofit were being done, more failures would have occurred – such as a carpenter who left air gaps between the floor and wall, as he thought the house would need more ventilation.

7. Miscalculations in efficiency modelling

A further cause of rebound effects is the miscalculations that often occur in planning and design. For example, a dwelling's orientation to the sun has a large influence on the amount of heat energy it will gain on an average winter's day, and this can be difficult to calculate due to shade from other buildings and the changing canopies of trees. It is also difficult to calculate heat losses that will occur through the interfaces of different components of the building envelope, such as window frames and walls; walls and the ground; and complex shaped bays and dormers. Further problems arise when calculations do not take technologies into account which interact with heating systems, such as supplementary fans to distribute warm air (Vázquez, 2013). It is also difficult to calculate the heat losses or gains that will occur between party walls in semi-detached, terraced and apartment properties (see discussion in Ingle *et al.*, 2014).

When defined as an elasticity (see Chapter 1), the rebound effect is a measure of the *change* in the ratio of calculated to actual heating consumption before and after a retrofit. This requires pre-retrofit parameters to be calculated as accurately as post-retrofit parameters. Many problems arise with this because often there is poor knowledge of the thermal performance of older materials and even some doubt as to what these materials are, especially if sandwiched between the inner and outer layers of walls (see discussion in Summerfield *et al.*, 2011).

8. Conclusions

Several factors can cause more heating energy to be consumed after a retrofit than would have been expected. The technology, the occupants' heating behaviour, and all the interactions between these, do not necessarily function as planned.

Rebound effects occur because households often operate their heating systems in a less disciplined way when they know the price of warmth is cheaper, and expand their household activities in space and time because their home is more comfortable throughout. Many households have difficulty inter-acting with newer, more technically complex thermostat and timing controls, and quickly learn to err on the side of over-heating rather than risk being cold. In some cases there are severe mismatches between heating technology and the daily or weekly rhythms of households' heating needs, and again many households err on the side of too much warmth rather than too little.

There can also be mismatches between the optimum output of heating systems and the most regularly occurring heating needs of occupants, causing boilers or heat pumps to run outside of their efficient ranges. Similar mis-matches can occur between the different pieces of equipment within a heating system, such as heat pumps and underfloor heating. Rebound effects can also occur due to construction workers' inadequate skill and know-how in exacting tasks such as applying wall insulation and forming joins between different parts of the building envelope. Poor calculations can raise expectations of perfor-mance, which are not realised in practice.

The term 'rebound effects' is used in this book to embrace these diverse factors. This is partly because the notion of a socio-technical system seems the best framework for understanding the manifold interactions between the human and thermal-technical elements of a home. This suggests that attempts to divide the causes of over-consumption into human and technical elements are misleading, and therefore a single figure for the rebound effect is appropri-ate. It also gives a straightforward figure which can assist policymakers, housing providers, homeowners and other stakeholders to plan their investments and priorities to achieve their goals. These stakeholders need to know what level of shortfall in energy saving is common or likely, so that planning can be adjusted accordingly.

Even if a retrofit goes entirely according to plan and there is no energy performance gap, i.e. consumption is within the level planned, there can still be large rebound effects. This is because a household that is consuming the same amount of energy as the calculated rating, in a perfectly retrofitted dwelling, may actually have been consuming little energy in comparison to the dwelling's calculated consumption rating before the retrofit – a phenomenon known as the 'prebound effect'. Prebound effects are, in fact, one of the most common reasons rebound effects happen after energy efficiency upgrades. This issue will be explored in the next chapter, both for its own sake and to lay the groundwork for a more detailed exploration of the rebound effect in Chapter 4.

3

THE PREBOUND EFFECT

1. Distinguishing 'prebound' from 'rebound'

The term 'rebound effect' emerged out of a stream of intense and insightful discussion among economists over the last two decades of the twentieth century, beginning with Daniel Khazzoom's ground-breaking paper in 1980 on the counter-intuitive effects of energy efficiency increases in household appliances (Khazzoom, 1980). The term 'prebound effect', however, was coined quite suddenly, during a research project in 2012, simply to give a name to a phenomenon which seemed something of the opposite of the rebound effect. It led to the publication of the article: 'Introducing the prebound effect: the gap between performance and actual energy consumption' (Sunikka-Blank and Galvin, 2012). Within just over a year the article had become the most ever downloaded from the journal in which it was published, *Building Research and Information*, a distinction it still held when this book was written. Two (unofficial) German translations of the article were also circulating, and it was used as the centrepiece of a popular television documentary on thermal retrofitting in Germany (Hanf, 2013).

The reason for this level of interest was that the prebound effect was a name given to a phenomenon that many researchers, building specialists, housing providers and municipal actors had already noticed, but few had realised they were not alone in this. Further, the phenomenon these people were observing had not been systematically quantified. It was of great concern to these people because it seemed to be severely compromising the amount of fuel savings being achieved through thermal retrofits of homes.

Sunikka-Blank and Galvin defined the *prebound* effect as a contrasting phenomenon to the *rebound* effect:

> The rebound effect is known to occur when a proportion of the energy savings after a retrofit is consumed by additional energy use, e.g. due to increased internal temperature and comfort expectations… By contrast, the 'prebound' effect refers to the situation before a retrofit, and indicates how much less energy is consumed than expected. As retrofits cannot save energy that is not actually being consumed, this has implications for the economic viability of thermal retrofits.
>
> (Sunikka-Blank and Galvin, 2012: 265)

Understanding this phenomenon is important for the creation of evidence-based policies and mechanisms. It is equally important for measuring the impact and efficacy of policies in practice. Once again, it shows that inhabitant behaviours play a significant role in delivering outcomes. Without a clear conceptual and numerical basis to account for this phenomenon, any policy is unlikely to deliver its intended results in energy and CO_2 reduction.

In most countries a figure is calculated to indicate the amount of heating energy a specific dwelling is expected to consume under average conditions of occupancy, indoor lifestyle and local weather. While this figure is called different things in different countries, and a variety of methods and assumptions is employed in calculating it (see e.g. Olfsten *et al.*, 2004), in this book it is given a general name, the 'energy performance rating' (EPR). In Europe the EPR is usually expressed in units of kilowatt-hours of energy consumed per square metre of floor area per year (kWh/m²a). Sometimes this is *final* energy, i.e. the energy consumed in the dwelling itself, and sometimes it is *primary* energy, which includes the energy that is wasted in bringing the energy source to the dwelling. Sometimes the 'floor area' over which this energy is consumed includes staircases, basements, corridors and lofts, particularly in multi-apartment buildings, but sometimes only the area inside a dwelling's front door is included. Care is therefore needed in interpreting EPR figures, but robust comparisons can be made as long as consistent definitions are used.

The phenomenon many had noticed occured in older homes that had not been significantly upgraded. The actual, measured heating consumption was very often significantly lower than the dwellings' EPRs. Therefore, when thermal retrofits were undertaken to reduce energy consumption, the achieved level of energy reductions were far smaller than anticipated. The reason for this is that households cannot make savings on energy they had not been previously consuming. This disappointed homeowners, frustrated energy planners, and raised questions about some countries' grounds for mandating certain levels of thermal retrofit standards. These mandates in Germany, for example, were based on the assumption that occupants who retrofitted their homes would save energy. However, it turned out they had not actually been consuming the assumed level of energy and had been more frugal (Galvin, 2014c).

The research by Sunikka-Blank and Galvin (2012) brought together EPR and consumption data from a number of other researchers, mostly in Germany but also in the Netherlands, Belgium and France. Existing German datasets from work by researchers such as Walberg *et al.* (2011), Loga *et al.* (2011), Erhorn (2007) and Jagnow and Wolf (2008) were sufficiently comprehensive to be able to be brought together, weighted to adjust for their different methodologies and numbers of dwellings, and synthesised to produce a preliminary estimate of how big the prebound effect was and how it varied with dwellings' EPRs. On this basis Sunikka-Blank and Galvin estimated that average actual heating consumption in German homes is about 35 per cent lower than dwellings' EPRs, and that this percentage gap widens as the EPR increases. They found very large differences, however, between individual dwellings, even within sets of physically identical houses. Some showed gaps of over 90 per cent, meaning that the occupants were consuming less than one-tenth the heating energy of their EPR. Others showed negative gaps, indicating they were consuming higher than their EPR. Most of the latter, however, were in modern, well-insulated homes. For homes built prior to 1977 – when insulation standards were first included in the building regulations – this gap averaged around 40 per cent, and increased to an average of about 60 per cent for very large, thermally poor homes.

The 'prebound effect' provided a simple metric for calculating average estimates of the energy savings that would not be possible through thermal retrofitting – because the occupants were already making these savings. In fact, the prebound effect is simply the numerical negative of the 'energy performance gap', which engineers frequently define as the excess energy consumption as a percentage of the calculated consumption (see Appendix, Section A.8). The metric of the energy performance gap had been widely used for comparing levels of *over*-consumption in new homes, and sometimes also in newly retrofitted homes. The metric of the prebound effect provided a measure of *under*-consumption in older existing homes, or homes prior to retrofitting.

The prebound effect is related to the rebound effect for a simple reason. If an older home is actually consuming less heating energy than expected, the household will not save as much energy as it expects to by retrofitting it. The increase in energy efficiency will not lead to a reduction in energy consumption in proportion to the energy upgrade. Instead, it leads to a smaller reduction, as a lower level of energy consumption existed before the energy efficiency upgrade occurred. It is the relationship between the pre-retrofit calculated consumption and the actual consumption that connects the prebound effect mathematically to the 'elasticity' rebound effect.

Work on the prebound effect has continued since the publication of the initial paper. This chapter brings an updated account of findings as they now stand, and how these relate to the elasticity rebound effect.

2. Estimating the prebound effect: synthesising the datasets

Over the last decade a number of German researchers have produced datasets of the EPRs and actual heating consumption of groups of homes. The number of dwellings in each study ranged from 44 to over 1,700, and together these gave a fairly full coverage of the types of older home in Germany and some coverage of newer or recently retrofitted homes. Table 3.1 gives the references of these studies, the numbers of dwellings investigated, their average EPRs, their measured consumption levels and their prebound effects. Energy figures in the datasets include both water and space heating.

Table 3.1 shows the average EPR of the dwellings in the datasets ranging from 175 to 261kWh/m^{2}a, while average measured consumption ranges from 135 to 170kWh/m^{2}a. This gives average prebound effects of 26–43 per cent. Weighting each dataset's average according to the number of dwellings in the dataset gives a weighted average prebound effect of 35 per cent. This implies that, for these 3,357 dwellings, average heating consumption is 35 per cent below the EPR. Households are consuming only 65 per cent of the officially calculated quantity of energy required to heat their homes to a comfortable level and provide adequate hot water.

A more nuanced indication of the prebound effect can be obtained by bringing the datasets together graphically, then weighting them according to the number of dwellings in each dataset to produce a synthesis graph of actual measured consumption against dwellings' EPRs. The initial graphs are displayed in Figure 3.1. This takes no account of the number of dwellings covered by each curve on the graph, but simply indicates how the curves lie. For example,

Table 3.1 Data sources and preliminary calculations for prebound effect in Germany. 'EFH' refers to detached houses (*Einfamilienhäuser*) and 'MFH' to dwellings in multi-dwelling buildings (Mehrfamilienhäuser).

Source	Type of data source	Type of dwellings	Number of dwellings	Average EPR (kWh/m²a)	Average measured consumption (kWh/m²a)	Average prebound effect (%)
Knissel and Loga (2006)	National random survey of 4,670 dwellings	In buildings with < 8 dwellings	1,178	261	150	43
		In buildings with 8+ dwellings	113	184	135	27
Loga *et al.* (2011)	National random sample	All types	1,702	220	152	31
Kassner *et al.* (2010)	Mixed sample, method unclear	All types	44	209	153	27
Jagnow and Wolf (2008)	Samples from OPTIMUS national survey	Not stated	Approximately 100	220	135	39
		Not stated	Approximately 100	200	148	26
Erhorn (2007)	Nationwide DENA study	Detached houses (EFH)	50	240	170	29
		Apartments (MFH)	70	175	140	30

Source: adapted from Sunikka-Blank and Galvin (2012)

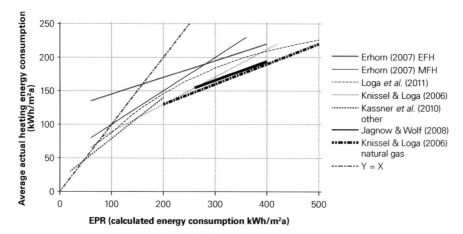

Figure 3.1
Relations between average actual heating energy consumption for specific values of calculated energy consumption

the plot for the data in Erhorn (2007) EFH (detached houses), which is significantly higher at the low-EPR end than the other curves, represents 70 detached houses, while the plot for Loga *et al.* (2011), which runs through the centre of the bulk of the curves, represents a nationwide random sample of 1,702 dwellings of all types. Hence in a synthesis of data the latter would be weighted more heavily than the former. The straight diagonal line labelled Y = X indicates where direct 1:1 correspondence between calculated and actual consumption would lie.

A synthesis of these curves, weighted according to the number of dwellings in each dataset, is given in Figure 3.2. The weighted average of all the datasets is given by the light continuous line, which is a fairly smooth curve apart from a raised portion in the range just below EPR 100kWh/m²a. The two other lines are formed by curve-fitting, i.e. finding a curve with a simple mathematical formula that closely approximates the original. The dotted line gives a very close fit, with a departure from the original of never more than 4.6 per cent and less than 2 per cent over most of the range. This curve has the formula:

$$E = 12.0C^{0.499} - 29.3 \qquad (3.1)$$

where E is the actual consumption and C the calculated consumption or EPR.

Although this is the best curve in terms of its fit with the data, a very close approximation is given by the heavier solid line on the graph, a power curve with the formula:

$$E = 4.5C^{0.641} \qquad (3.2)$$

This curve has slightly larger differences with the actual data than the previous curve, but is still a good representation, given that the differences are mostly

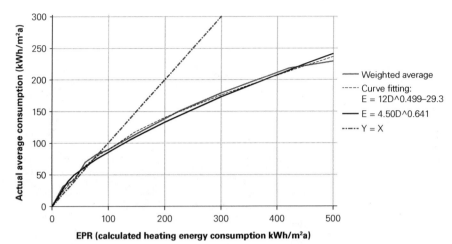

Figure 3.2
Weighted average of data points, with curve-fitting

under 2 per cent and range above 5 per cent only at the extreme ends of the curve. This formula is much more flexible to use in mathematical analysis of the prebound and rebound effects, as it has only one term on the right hand side. As will be shown, this curve indicates there is a consistent, average rebound effect in these dwellings of around 36 per cent (subtract the exponent 0.641 from 1, to get 0.359, and multiply it by 100 to turn it into a percentage). The mathematical proof for this is given in the Appendix, Section A.4.

The prebound effect was defined in Sunikka-Blank and Galvin (2012) as the proportionate or percentage gap between actual consumption and EPR, with respect to the EPR. A little manipulation of equation (3.2) (see Appendix, Section A.10) shows that the prebound effect can therefore be expressed as:

$$P = 1 - 4.5C^{-0.359} \tag{3.3}$$

This curve is displayed in Figure 3.3. It shows a prebound effect of over 50 per cent for dwellings with an EPR above 450kWh/m²a, reducing to 35 per cent for dwellings with EPR 220kWh/m²a, the average EPR for Germany (DENA, 2012). The prebound effect reduces further for dwellings with lower EPRs, i.e. more energy efficient dwellings, reaching zero at EPR 70kWh/m²a and becoming negative for yet more energy efficient dwellings.

This does not mean that every dwelling's prebound effect falls precisely on the line of the graph in Figure 3.3. Rather, each point on the curve represents the *average* prebound effect for dwellings with an EPR of that value. At any particular EPR there is a very large range of prebound effects, and the value for any particular household depends largely on occupants' heating behaviour, the number of occupants, the size and layout of the house, and the accuracy of the EPR calculations. Generally, though, if a random sample of German dwellings of any particular EPR is examined, the prebound effect can be expected to be close to that displayed on the graph.

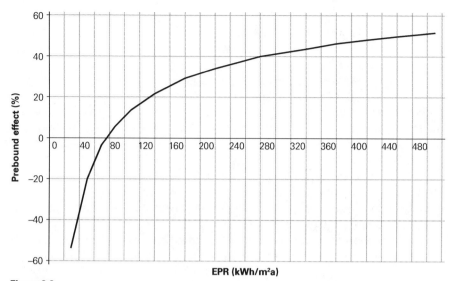

Figure 3.3
Average prebound effect for German dwellings, based on combined dataset

3. Comparisons with other countries
Similar or comparative datasets to those above are gradually being produced for at least four other countries.

3.1 UK
The UK uses an alternative metric to the EPR, namely the 'standard assessment procedure' (SAP) rating. The SAP is on a scale of 0–100 and works in reverse to the EPR: the lower the EPR (i.e. the more thermally efficient a dwelling is), the higher the SAP rating. It is difficult to compare the two directly, as the curvature of the scales is different (see information on this and recent updates in English Housing Survey, 2012). UK researchers used SAP data and actual consumption data from the English House Condition Survey[1] of 1996 to estimate the gap between actual and calculated consumption in the UK housing stock (Laurent et al., 2013). The survey covered a statistically representative sample of UK housing, so its results are likely to be representative for homes at that time.

Average actual consumption was 72 per cent of average calculated consumption based on the SAP rating, which represents an average prebound effect of 28 per cent. Similar to the German situation, the prebound effect increased as the EPR increased (i.e. as the SAP reduced). For homes with the highest SAP rating the prebound effect was around 5 per cent, while for those of the lowest SAP rating it was 55 per cent.

Actual compared to calculated consumption was also investigated in relation to the age of the dwellings. Dwellings built prior to 1850 showed an average prebound effect of 41 per cent. Those built between 1950 and 1980 showed a fairly consistent average prebound effect of 23 per cent, while for those built after 1980 this narrowed to approximately 16 per cent.

3.2 The Netherlands

The Netherlands' Ministry of Housing conducts a survey of dwellings each year known as the 'WoON' survey, an abbreviation for *Het wonen overwogen*, the Dutch for 'Considering accommodation'. This survey included 4,700 households, and results where weighted according to house types to improve the sample's representativeness for the country. As part of the survey an EPR was calculated for each dwelling and actual consumption was recorded. The raw data was not made available to the public, but a large number of summarised findings are presented in Tigchelaar and Leidelmeijer (2013), and further in Tigchelaar (2011) and Laurent *et al.* (2013).

A convenient way to display the data is in terms of average prebound effects for each class of energy label. This is given in Figure 3.4. An 'A' label is the most energy efficient, with the lowest EPR, while the 'G' label is the least efficient and has the highest EPR.

As for the German case, the prebound effect rises as the EPR increases. It is noteworthy, however, that the average rebound effect in each class does not reach 50 per cent, even for the least energy efficient homes, and does not fall below 12 per cent for 'A'-rated dwellings. Nevertheless, the *range* of prebound effects within each band is large. Five per cent of 'A'-rated dwellings have prebound effects lower than −30 per cent, while 5 per cent of 'G'-rated dwellings have prebound effects above 79 per cent.

3.3 France

Cayla *et al.* (2010) conducted a questionnaire survey of 2,012 French households in June 2009, including questions on the technical features of their homes, their consumption practices and fuel bills. Questionnaires which were incorrectly or incompletely filled in were discarded, and the results of the remaining 913 were weighted to reflect the proportions of housing types, heating fuels, income groups, etc., for France as a whole. The heating portions

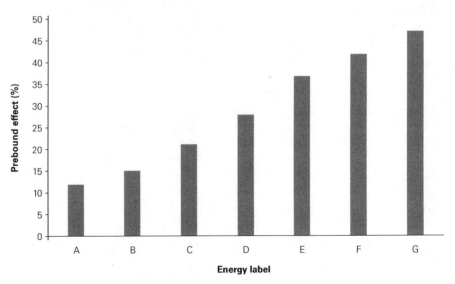

Figure 3.4
Prebound effect in Dutch dwellings
Data source: Tigchelaar (2011), own calculations

of the bills were separated from the remainder of the energy bills using a method developed by the French research centre CEREN (http://ceren.fr/index. aspx), and the results were checked for consistency with proportions nationwide.

The researchers were thereby able to offer reliable estimates of calculated and actual heating consumption for these 913 households.[2] A plot of these figures is given in Figure 3.5. Dwellings with EPRs above 500kWh/m²a are omitted from the graph (there are few of these, with extremely high calculated consumption values, and there inclusion makes the graph less intuitively interpretable) but are included in the calculations for equations (3.4) and (3.5).

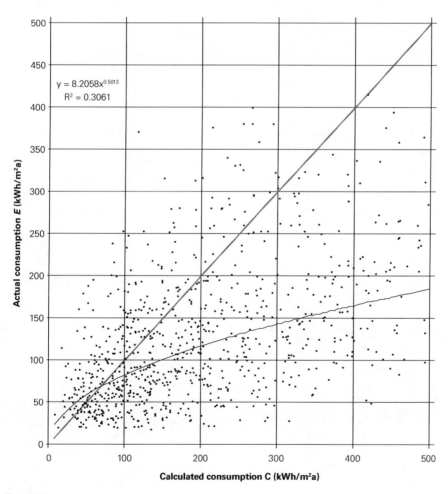

Figure 3.5
French data: estimated space heating consumption as a function of normative calculation, 913 homes
Data source: Cayla et al. *(2010), own calculations*

The straight diagonal line in Figure 3.5 represents the points where calculated and actual consumption would be equal. The curved line is the best fit power curve:

$$E = 8.2058C^{0.5013} \qquad (3.4)$$

This indicates a prebound effect of:

$$P = 1 - 8.2058C^{-0.4987} \qquad (3.5)$$

This is displayed in Figure 3.6. It is interesting to compare this with the German case. Both show prebound effects of zero where the EPR is around 70kWh/m²a, becoming deeply negative (i.e. having high energy performance gaps) for EPRs below that level. However, in the German case the prebound effect remains below 60 per cent even for homes with very high EPRs, whereas for the French dataset the prebound effect edges over 60 per cent for EPRs over 430kWh/m²a. This indicates that these dwellings are, on average, consuming less than half the quantity of heating energy that was calculated for dwellings of their type, size and thermal characteristics.

3.4. Belgium

While no direct comparison of actual and calculated consumption is available for Belgium, Hens *et al.* (2010) conducted a study with similar indicators, for 924 Belgian homes. This compared annual actual heating consumption with a metric which represents the thermal quality of the building, namely the average U-value of the building envelope U divided by its volume V. There is a direct relationship between a building's volume and its ability to retain heat. A large

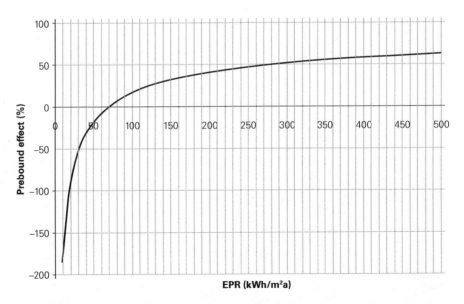

Figure 3.6
Prebound effect in 913 French dwellings
Data source: Cayla et al. (2010), own calculations

building has less surface area per unit volume than a small building, and therefore loses heat more slowly in proportion to its size. Hence the factor U/V is a good proxy for the thermal leakiness of a building, and is likely to be proportional to the EPR. The relationship between this factor and actual heating consumption that Hens and colleagues found for these buildings was:

$$E = 299.6 \times \left[\frac{U}{V} \right]^{0.84} \tag{3.6}$$

Again this has the form of a power curve, showing that the gap between actual and calculated consumption increases as the factor U/V increases. Hence, as with the results for the other four countries, the prebound effect is larger for less energy efficient buildings. However, the relatively high exponent 0.84 suggests that the prebound effect in these dwellings is significantly lower than in those of the other countries. The reason might be that the Belgian formula does not take into account the quality of the boiler and heating system, nor the air-tightness of the buildings. More research would be needed to find what sort of difference these factors make to the results.

In general, however, the prebound effect is clearly evidenced in datasets for all countries where the relevant parameters have been investigated in reasonably large samples of dwellings. It may well be a universal phenomenon.

4. How the prebound affects energy saving

The post-retrofit performance of three apartment blocks in Germany was described in Chapter 2. The apartments with traditional retrofit technology were designed to consume 50kWh/m²a, and after two years of operation their actual consumption was 51kWh/m²a (Galvin, 2014a). This seemed like a very successful undertaking, as it appeared that almost all the energy that was expected to be saved, was indeed saved. However, this does not take into account the prebound effect in the apartments prior to the retrofit. The pre-retrofit EPR of this apartment block was 320kWh/m²a, while the actual, measured consumption was 171kWh/m²a, a prebound effect of 46.6 per cent. This is close to the average prebound effect for this EPR, as displayed in Figure 3.2, of 43.3 per cent.

If the definition of the rebound effect as the 'energy performance gap' is used, this apartment block showed a rebound effect of only 2 per cent after retrofitting, as this definition does not take the prebound effect into account. However, when the definition of the rebound effect as an 'elasticity' is used, the prebound effect is taken into account in the mathematics, and this reveals the rebound effect to be much larger, indeed just over 35 per cent.

Prebound effects are not taken into account in the calculations used in Germany to estimate fuel savings under the Energy Savings Regulations (EnEV, 2009). These are based strictly on the EPRs before and after retrofitting. It was expected that the retrofitted apartment block noted above would save 270kWh/m² per year, but in fact it saved only 120kWh/m² per year. With a total floor area of 2,100m², the savings expected for the building were 567,000kWh per year, while the actual savings were 252,000kWh. Over half the expected savings failed to materialise.

This narrative can be extended to a national level. Every year in Germany approximately 0.85 per cent of the housing stock undergoes thermal retrofitting, and the average EPR in these homes is reduced by about 50 per cent (Galvin and Sunikka-Blank, 2014b). This would imply a reduction from the average EPR of 230kWh/m²a to around 115kWh/m²a, a saving of 115kWh/m²a. Since the average floor area in older homes is 85m² and there are 38 million occupied dwellings in Germany, this implies that the expected annual energy savings from these retrofits is around 3.16TWh (1 TWh = 1,000 million kWh).

However, with the prebound effect taken into account, the actual consumption before retrofitting is closer to 150kWh/m²a, which is in fact the average actual consumption of the German building stock (Schröder *et al.*, 2011; Walberg *et al.*, 2011). If the retrofits achieve their target EPR of 115kWh/m²a, their actual consumption will now be approximately 95kWh/m²a (see Figure 3.2). Hence the actual saving will be 55kWh/m²a, or 37 per cent of the original actual consumption. This equates to actual annual energy savings for Germany of 1.51TWh, about half what was expected. This is a significant shortfall, which can frustrate national energy policy goals, as well as leaving occupants worse off than expected.

This calculation covers only the homes with substantial energy efficiency improvements to their building envelopes, including new windows and wall, floor and ceiling insulation. It does not take into account a large number of other homes where just one or two windows are replaced, nor does it cover the replacement of boilers and heating systems. When these are taken into account the same pattern emerges (Galvin and Sunikka-Blank, 2014b). Only about half the expected quantity of energy is saved.

Retrofit policy in the UK has made attempts to take this issue into account over the last decade, and there are now efforts to account for it in research projects in the US and Germany. A recent project in the US used computer modelling of a sample of 322 households, based on assumptions about the actual consumption characteristics of these types of household (Ingle *et al.*, 2014). This was not straightforward, as there is a great variety of household consumption behaviours even among apparently similar households (Gram-Hanssen, 2013). In the US sample in homes heated by natural gas, average actual consumption was 18 per cent lower than average EPR,[3] and 4 per cent lower for homes heated with electricity – i.e. prebound effects of 18 per cent and 4 per cent. The computer models could partially estimate these prebound effects, but did not fully account for it. The computers' estimated prebound effect figures were 11 per cent and 3 per cent respectively, the first of which is quite a significant underestimate.

The computer models showed that the energy saving potential of these households was about 15 per cent less than would be expected if based on their EPRs. The actual consumption figures show that this shortfall would have been 40–50 per cent. This attempt at modelling is a useful start, but clearly needs improvements if it is to take full account of likely prebound effects.

A less intricate modelling method for estimating the gap between actual consumption and EPR was attempted in Germany (Knissel and Loga, 2006). The motivation for this study was that rents in Germany can be influenced by the thermal quality of a dwelling, i.e. the EPR. For old buildings, however, it is expensive and technically demanding to calculate the EPR, due to uncertainties in the types of building material. By comparing actual consumption with theoretical consumption in buildings where both of these were known, the authors of the study devised a formula for estimating the EPR of buildings of a wide range of classes, based on known actual consumption. This gave the reverse information to that in the US study, which set out to estimate actual consumption: it provided a rough guide for calculating the EPR if the actual consumption was known. The authors claimed to be able to estimate the EPR to an accuracy of within 10 per cent using this method. Nevertheless, there do not seem to be any moves to use a method such as this in reverse on a large scale in Germany, to estimate actual consumption and use this to predict the actual energy savings that would be likely after a retrofit. Instead, the Energy Savings Regulations maintain their emphasis on estimating savings based on EPR alone, and the EPR is still calculated using only technical information about the building itself.

In the UK there have been attempts over the last decade to take account of occupant behaviour in estimating likely savings due to a retrofit. A recent example is for the 'Green Deal', a government initiative to provide loans for certain retrofit measures. To qualify for a loan, the expected reductions in a household's monthly energy bill due to the retrofit measures have to cover the cost of the loan repayments. A calculation of the costs and savings is required for each specific household: this entails more than just estimating the EPR before and after the retrofit. Hence a two-part estimate of pre-retrofit consumption has been devised. First, a technical investigation leads to an 'Energy Performance Certificate' (EPC). This rates the property in terms of energy efficiency, i.e. it produces an EPR, and gives recommendations for improvement. Second, an 'occupancy assessment' modifies the output of the EPR to reflect a household's actual energy use (DECC, 2014a). The occupancy assessment takes into account the number of persons in the household, their energy bills over the last 12 months, their reported occupancy patterns, and their patterns of usage of appliances such as showers and washing machines.

At the time of writing, only two cases are known to this author where the results of this two-pronged estimate could be compared to actual consumption. The first case is a three-bedroom semi-detached house in Cambridge, UK. The EPR estimated heating consumption at $200kWh/m^2a$, and this estimate was reduced to $144kWh/m^2a$ when the occupancy assessment was taken into account, an estimated prebound effect of 28 per cent. However the actual, measured consumption over the previous two years was $69kWh/m^2a$ – an actual prebound effect of 66 per cent. In an interview with researchers the householder pointed out that the energy advisor's software was not able to accept the nuances of the household's heating patterns. The household had devised a method of 'zonal' heating, where only the rooms actually being used were heated, and only while they were being used. Hence it was impossible to provide the computer program with a realistic estimate of the long-run average indoor temperature of the entire interior of the house, and it defaulted to the average temperature in the lounge.

The second case was a large semi-detached Victorian house, also in Cambridge, which had already been comprehensively retrofitted. The owners were considering further energy efficiency measures. In this case the Green Deal software estimated actual heating consumption very accurately, at 70kWh/

m²a. A major difference here was that the occupants use an air heating system which keeps all rooms constantly at the same target temperature. It appears the software performs well when there are no non-standard aspects to the occupants' heating behaviour.

If inaccuracies are noted and overcome it might be possible over time to correct these and improve software packages.

It is important for policymakers, the housing industry, homeowners and other stakeholders to be able to make reliable estimates of the likely energy savings from energy efficiency upgrades of various types on specific houses, as well as on the housing stock as a whole. The prebound effect identifies a key reason these estimates very often go wrong. Where this is recognised and accepted, progress is being made to devise methods of improving the accuracy of predictions.

5. Prebound and rebound interactions

As noted above, the prebound effect is the numerical negative of the 'energy performance gap', which, it was noted in Chapter 1, serves as one definition of the rebound effect. For example, if a dwelling has a prebound effect of −20 per cent (which, as Figure 3.3 suggests, would be the case with an average German dwelling of EPR 40kWh/m²a), this dwelling has an energy performance gap of +20 per cent. It is consuming 20 per cent more energy than it was designed to consume, for what is regarded as a normal occupation and heating pattern. However, the energy performance gap suffers limitations as a definition of the rebound effect. Some of these are overcome when we consider how the prebound effect relates to the 'elasticity' rebound effect, as explained below.

The prebound effect also can be used to calculate the elasticity rebound effect. This gives a measure of how energy *services* consumption *changes* after an energy efficiency improvement. Energy services are the *benefits* households get from consuming energy. These benefits are such things as higher indoor temperatures, longer heating periods and better indoor air quality. It is not easy to measure energy services in domestic heating, as different benefits can cancel each other out or add to each other: one household may heat more rooms for longer periods but to a lower temperature, while another may heat fewer rooms for shorter periods but to a higher temperature. Even if these benefits could be measured on a robust quantitative scale, there are no agreed units (such as kilowatt-hours or calories) for energy services.

Engineers therefore use a proxy for energy services, namely the ratio of actual consumption to EPR (see discussion in Galvin, 2014a). For example, if a house has an EPR of 200kWh/m²a but its occupants consume only 100kWh/m²a, they are said to be taking energy services of 0.5. If they consume 200kWh/m²a in such a house, they are said to be taking full energy services, or energy services of 1.0. Usually, after a retrofit the level of energy services increases. This increase is the essential feature of the elasticity rebound effect.

In the Appendix, Section A.10 shows how the prebound effect can be used to calculate the elasticity rebound effect. The prebound effect for the French housing stock was seen above to be represented by the formula:

$$P = 1 - 8.2058C^{-0.4987}$$

(3.7)

The exponent in this expression, −0.4987, is the numerical negative of the elasticity rebound effect. Hence the elasticity rebound effect for this French dataset is 0.4987, or 49.87 per cent. What this means in practical terms will be discussed in more detail in Chapter 4. Meanwhile it is important to note that there is a direct mathematical link between the prebound effect and the elasticity rebound effect.

6. Conclusions

The notion of the 'rebound effect' was developed through intense discussion and research among economists and engineers during the last two decades of the twentieth century. It provides a set of definitions and tools for conceptualising and quantifying the shortfalls in the energy saving which frequently occur after energy efficiency increases. It is a flexible and robust concept which can be used in all energy consumption sectors, such as housing, transport, industry and services. The prebound effect is a much narrower and more limited concept, which was developed to fill a gap in studies of rebound effects in domestic heating. It gives a measure of how much less energy a household in a non-energy efficient dwelling is consuming, in comparison to that dwelling's EPR.

Prebound effects have been observed and measured in Germany, the UK, the Netherlands, Belgium and France, and are widely acknowledged to occur in all countries where there is research on energy efficiency upgrades in domestic heating. Extensive research on the German housing stock shows average prebound effects of around 35 per cent, rising to over 40 per cent for older homes and over 50 per cent for the least energy efficient. For France the figures seem to be higher, reaching over 60 per cent for the least energy efficient. These observations suggest that the prebound effect is significant and impacts upon energy and retrofit policies. Both public policy and practice will need to account for and address this.

A simple formula can been derived to predict average prebound effects, where a representative sample of actual consumption and EPR of buildings is available in any particular country or housing stock.

The prebound effect has a direct relationship with the 'rebound effect' as defined as an 'energy performance gap': it is simply the numerical negative of this. Its relationship with the 'elasticity' rebound effect is more subtle: it is the numerical negative of the exponent in the prebound effect formula. The significance of the values obtained in these calculations will become clear as the rebound effect is considered in more detail in the following chapter.

Notes

1   www.gov.uk/government/collections/english-housing-survey
2   The raw data used in Figure 3.5 is not given in Cayla *et al.* (2010) but was kindly given by these authors for use in this book. This author is grateful for the generosity of Benoit Allibe, Marie-Helen Laurent and Jean-Narie Cayla of Électricité de France, for making this data available for use in this book.
3   Ingle *et al.* (2014) express this in reverse, i.e. the EPR was 22 per cent higher than actual consumption, while their figures for modelled consumption are expressed as a percentage lower than the EPR. The method is made consistent here to enable direct comparisons to be made.

4

METHODS FOR ESTIMATING THE REBOUND EFFECT IN DOMESTIC ENERGY CONSUMPTION

1. Introduction

Economists led discussion of the rebound effect in the 1980s and 1990s, and by the end of last century their definition of the rebound effect as an 'elasticity' had become dominant. In this definition the rebound effect is the ratio between the proportionate change in the consumption of energy services and the proportionate change in energy efficiency (see Chapters 1 and 3).

This definition has four main advantages over others that have surfaced. First, it can be adapted to any sector of the economy, making it possible to compare rebound effects coherently in such fields as housing, surface transport, air transport, heavy industry and services. Second, it is mathematically robust and adaptable, as it is based on the branch of mathematics known as calculus, which focuses on infinitesimally small changes in quantities such as time, distance, wealth, economic output, energy and energy efficiency. This enables it to estimate rebound effects when these quantities change in linear or non-linear ways, instantly or over periods of time. Third, it enables mathematical models to be developed which closely approximate what actually happens in real life with energy and economics. This is because many changes that take place from year to year, such as wage or price increases, are proportionate to the previous year's absolute value – such as where prices rise by 2 per cent per year. The mathematics of elasticities is ideally suited to dealing with this kind of 'growth function'.

Fourth, the concept of an elasticity is used more widely in economics than merely in relation to the rebound effect, and this enables comparisons between rebound effects and other effects to be made. The 'price elasticity of consumption' is the ratio between the proportionate change in the consumption of a good, such as energy, and the proportionate change in the price of that good. If the price elasticity of energy consumption is known, this can sometimes be used to infer the rebound effect, via a series of mathematical steps (see, for example, Madlener and Hauertmann, 2011). Even where this is not possible, human and societal responses to changes in various quantities of goods can be usefully compared to rebound effects using the common concept of an elasticity.

In relation to domestic heating, two further definitions of the rebound effect have emerged and been found useful in situations where there is not enough information to calculate an 'elasticity' rebound effect. One of these, described in Chapters 1 and 3, is defined in this book as the 'energy performance gap' (EPG): the ratio between over-consumption and expected or calculated consumption. This is especially useful where there have been no *changes* in consumption or efficiency, such as with new housing, and only two items of information are known: the calculated consumption and the actual consumption. As will be seen below, this is a completely different measure from the elasticity rebound effect, and if both are used in the same case study they can give values that are widely divergent.

A third definition emerged in an important study published at the turn of this century (Haas and Biermayr, 2000), for situations where three items of information are known: the actual consumption before and after an energy efficiency upgrade, and the calculated consumption after the upgrade. In this book, this definition is entitled the 'energy savings deficit' (ESD) as it offers a numerical estimate of the proportion by which the energy saved, through an energy efficiency upgrade, falls short of what was expected or calculated. Again, this is a very different measure from both the others and, as will be shown below, its results in the same case study can be very different from that of the other definitions of the rebound effect.

In a simple case study of a dwelling being thermally upgraded, calculating the elasticity rebound effect requires four items of information: the actual *and* calculated consumption before *and* after an energy efficiency upgrade. The most frequently missing of these is the calculated consumption prior to the upgrade, which is why the ESD is often used as a next best option. This has created the problem that many studies of rebound effects in domestic heating give values which are widely divergent, even though, in fact, the realities may be much more similar than the figures suggest. A table in Galvin (2014a) gives a list of 17 such studies – a selection taken from a much larger number – showing how divergent their rebound effect definitions are. One of the studies in this list, Yun *et al.* (2013), includes its own summary of studies and their divergent definitions and results.

The elasticity rebound effect cannot be calculated for a case study dwelling unless all four of the data items are known. There are two other situations, however, where it can be calculated if only two of these are known. These are cross-sectional methods and time series methods. It can also be calculated in what are known as 'proxy' methods, where some parameter other than energy efficiency is used as a stand-in. The remainder of this chapter describes these and the case study method of estimating rebound effects.

2. Case study method

The case study method of estimating the rebound effect is illustrated here by two dwellings in Germany which were recently thermally upgraded. One is called here the 'Nuremberg case' (see Galvin, 2014a) and the other the 'Potsdam' case (as yet unpublished). Due to data protection obligations, more precise locations of these homes cannot be given. Table 4.1 gives the four data items for each of these upgrades.

Table 4.1 Calculated and actual consumption (kWh/m²a primary energy) before and after energy efficiency upgrades, for two case study dwellings in Germany.

Case study building	Before upgrade		After upgrade	
	Calculated consumption (kWh/m²a) C_1	Actual consumption (kWh/m²a) E_1	Calculated consumption (kWh/m²a) C_2	Actual consumption (kWh/m²a) E_2
Nuremberg	275	170	97	102
Potsdam	235	102	121	58

2.1 Nuremberg case study

A schematic of the consumption values for the Nuremberg case study is shown in Figure 4.1. The horizontal axis represents calculated consumption C and the vertical axis actual consumption E. The point labelled 'before' marks where E and C lay prior to the upgrade, while 'after' gives their values after the upgrade. The line marked $E = C$ is where these points would lie if the actual consumption were always equal to the calculated consumption. This case is highly typical of most of the energy efficiency upgrade cases this author has seen, in that the 'before' point lies below the line $E = C$ while 'after' lies above it. This means that, prior to the upgrade the household was consuming less energy than the calculated consumption as recorded in the dwelling's energy performance rating (EPR), and after the upgrade it was consuming more than the new EPR.

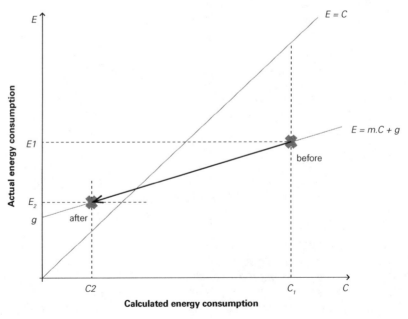

Figure 4.1
Schematic of calculated and actual consumption for Nuremberg case study dwelling

Of the three definitions of the rebound effect considered here, the *EPG* is the easiest to calculate (see Appendix, Section A.8), as it considers only the post-upgrade actual and calculated consumption:

$$EPG = \frac{E_2 - C_2}{C_2} = \frac{102 - 97}{97} = 0.052 \text{ or } 5.2\%$$ (4.1)

This result suggests the upgrade has been very successful, as the 'rebound effect' by this definition is the low value of 5.2 per cent.

The next definition, the *ESD*, takes the pre-retrofit actual consumption into account (see Appendix, Section A.9). It is obtained by dividing the shortfall in energy savings, by the 'expected' savings, i.e. the difference between the *actual* consumption *prior* to the upgrade and the *calculated* consumption *after* the upgrade:

$$ESD = \frac{E_2 - C_2}{E_1 - C_2} = \frac{102 - 97}{170 - 97} = 0.68 \text{ or } 6.8\%$$ (4.2)

This result, too, suggests the upgrade has been very successful, as the 'rebound effect' by this definition is a mere 6.8 per cent.

The problem with these two definitions, however, is that they do not reveal how the consumption behaviour of the household has changed as a result of the upgrade. Looking again at the figures in Table 4.1, it appears that prior to the retrofit this household was very frugal with heating energy. In a dwelling that required 275kWh/m²a to be fully heated, the household consumed only 170kWh/m²a. Their energy *services* consumption amounted to 170/275 = 0.62 of a fully heated home. After the upgrade, however, they consumed 102kWh/m²a, in a dwelling that required only 97kWh/m²a to be fully heated. Their energy services consumption had risen to 102/97 = 1.05 times that of a fully heated home. They had increased their consumption of energy services by 69 per cent of its original level.

The change in energy services is one of the parameters used in calculating the elasticity rebound effect. The other is the change in energy efficiency. As was noted in Chapter 1, there is no universal unit for energy efficiency of dwellings, but the reciprocal of the EPR gives a useful comparative figure (if the EPR is halved, then the energy efficiency is doubled). In this case, the energy efficiency changed from 1/275 = 0.00364 to 1/97 = 0.01030, an increase of 183 per cent.

If this household kept the same indoor temperatures and heating times after the upgrade as before, their increase in energy services consumption would be 0 per cent, and the elasticity rebound effect would also be zero. In fact, the elasticity rebound effect in this case is 51 per cent. The formula used to calculate this is given below. The mathematical derivations behind this are given in the Appendix, Section A.5. The symbol 'ln' means 'the natural logarithm of'.

$$R = 1 - \frac{\ln\left(\dfrac{E_1}{E_2}\right)}{\ln\left(\dfrac{C_1}{C_2}\right)} = 1 - \frac{\ln\left(\dfrac{170}{102}\right)}{\ln\left(\dfrac{275}{97}\right)} = 0.510 \; or \; 51.0\% \qquad (4.3)$$

This indicates that 51 per cent of the increase in energy efficiency has been used to increase the consumption of energy services, while the remaining 49 per cent has gone to reducing energy consumption. Seen from the perspective of the elasticity rebound effect, this upgrade project looks much less successful than it does if only the EPG or the ESD is considered. This is because the elasticity rebound effect takes all four of the significant factors into account in estimating the changes in consumption which result from an energy efficiency upgrade. It thereby gives a very clear indication of how much more energy service consumption this household indulged in, after an energy efficiency upgrade of that size. The three results are displayed in Figure 4.2.

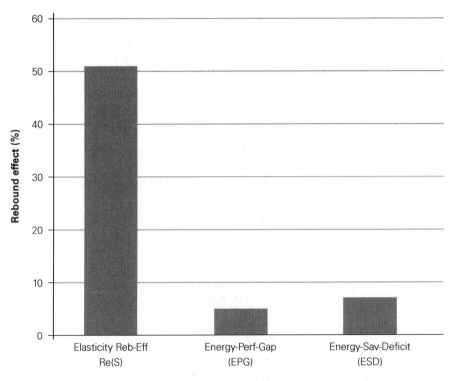

Figure 4.2
Rebound effect: the three definitions for the Nuremberg case study building

2.2 Potsdam case study

Figure 4.3 gives a schematic of the consumption values for the Potsdam case study dwelling. There are two notable differences between this and the Nuremberg case. First, post-upgrade actual consumption is lower than post-upgrade calculated consumption. This makes the EPG negative:

$$EPG = \frac{E_2 - C_2}{C_2} = \frac{58 - 121}{121} = -0.521 \, or -52.1\% \tag{4.4}$$

This would be regarded as a very good result, if the EPG were the only metric being considered: the home is consuming 52.1 per cent less energy than it was designed to. Second, however, unlike the Nuremberg case, the pre-upgrade *actual* consumption was *lower* than the post-upgrade *calculated* consumption: the occupants were already consuming less energy than the upgrade was aiming for. This gives a nonsensical result for the ESD:

$$ESD = \frac{E_2 - C_2}{E_1 - C_2} = \frac{58 - 121}{102 - 121} = 3.316 \, or \, 331.6\% \tag{4.5}$$

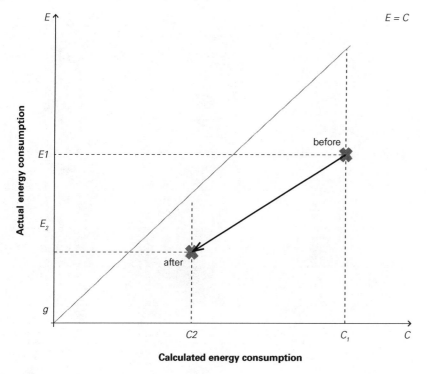

Figure 4.3
Schematic of calculated and actual consumption for Potsdam case study dwelling

This is an extremely high value, but it has no relation to reality because this metric was not designed to take account of cases where occupants consumed less energy prior to an upgrade, than the upgrade was aiming for. Comparing these two results of −52.1 per cent and 331.6 per cent gives an indication of how important it is to make definitions of the rebound effect clear, and to choose definitions that are appropriate to the situation, when estimating rebound effects for the benefit of policymakers and other stakeholders.

These problems are avoided when the elasticity rebound effect is used. Here the result is:

$$R = 1 - \frac{\ln\left(\dfrac{E_1}{E_2}\right)}{\ln\left(\dfrac{C_1}{C_2}\right)} = 1 - \frac{\ln\left(\dfrac{102}{58}\right)}{\ln\left(\dfrac{235}{121}\right)} = 0.150 \; or \; 15.0\% \tag{4.6}$$

This means that 15 per cent of the energy efficiency upgrade is used to increase the take of energy services, while the remaining 85 per cent goes to reduce energy consumption. For energy planners this is a better result than that for the Nuremberg case, as it indicates that the energy efficiency upgrade has been used more effectively. Figure 4.4 displays the results for the three definitions.

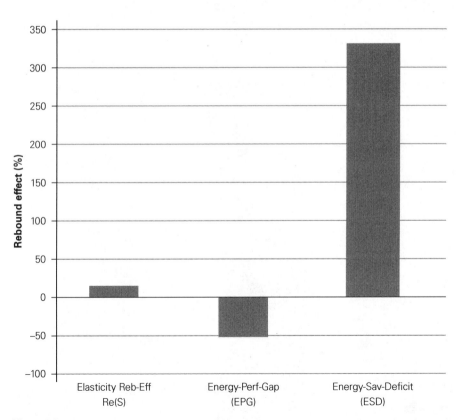

Figure 4.4
Rebound effect: the three definitions for the Potsdam case study building

A number of other case studies, each using the three definitions of the rebound effect, are presented in Galvin (2014a). Of the three definitions, the elasticity rebound effect is the only one that produces consistently coherent results. If either of the other definitions is used in research reports, this must be made very clear, not only in the reports themselves, but also in summaries and reviews of studies on the rebound effect which are very often produced by government and international agencies (e.g. DECC, 2008; EST, 2013; Maxwell and McAndrew, 2011).

3. Time series method

3.1 Changing energy consumption and efficiency overtime
If it is known how the average energy efficiency and average heating energy consumption of a country's housing stock have changed over a sufficiently large number of years, then a value can be calculated for the rebound effect. For example, the average energy efficiency of the UK housing stock increased by 1.81 per cent per year in 2000–2011, while the average energy consumption reduced by 1.43 per cent per year in that period (based on data in Odyssee, 2013). As was shown in Chapter 2, this implies an average rebound effect of just under 20 per cent. This means that, for every 1 per cent increase in energy efficiency throughout this period, the consumption of energy services increased by 0.2 per cent, while energy consumption reduced by 0.8 per cent.

There are, nevertheless, two important caveats with this result and the others that will be shown below. First, it is by no means certain that this steady, year-by-year reduction in energy consumption – which is evidenced in most European housing stocks – has been due entirely to the steady improvement in energy efficiency. There is strong evidence from European countries, particularly Germany, that only about half the year by year reduction in heating energy consumption can be explained by increases in energy efficiency (Galvin and Sunikka-Blank, 2014b). It seems that households are generally heating more frugally. This is possibly due to spending less time at home and more at work, and possibly because there are increasing numbers of empty bedrooms due to the trend to live alone or in smaller households. This would make these rebound effect results significantly lower than results based entirely on responses to changes in energy efficiency.

The second caveat is that the figures used for energy consumption and efficiency here include not only heating, but also electrical appliances, cooking, etc. Space and water heating accounted for 82 per cent of EU households' energy consumption in 2000 and 79 per cent in 2011, while energy efficiency reduced at a lower rate for heating than for other household energy consumption (ENERDATA, 2014). This would put upward pressure on rebound effect results.

Hence it is unlikely that these results will give a true indication of the rebound effect for domestic heating. However, they do provide another important function. They enable policymakers and other stakeholders to see at a glance what numerical increase or decrease in domestic energy (and energy services) consumption is occurring, in relation to the energy efficiency gains that are being achieved. They show, for example, that policymakers will not necessarily reduce a national housing stock's energy consumption by a certain percentage simply by increasing its energy efficiency by that percentage. Instead, these results give actual, country by country figures which can be used to estimate the amount by which energy efficiency will need to increase, in order for specific energy consumption goals to be met – assuming that current trends in consumption with respect to energy efficiency continue. This provides useful feedback to policymakers (and others) on whether and how specific policies are delivering their anticipated results.

3.2 Time series method

The EU Commission collects data on the energy efficiency and energy consumption of housing stocks in all 28 EU countries plus Norway, and makes this publicly available in its Odyssee (2013) database. A further publication explains the collection of the data and discusses its reliability (ENERDATA, 2012). The raw data points for each of the years 2000–2011 can be processed to give values for rebound effects in each country, subject to the caveats noted above. The steps in processing are as follows.

First, the raw data for each country has to be smoothed so that it can be modelled by a simple mathematical function which can readily be used in elasticity calculations. Taking Portugal as an example, the continuous line in Figure 4.5 shows how the actual data points for energy efficiency link up. Since values for energy efficiency are relative to any chosen datum, the data has been indexed to the value 1 for the year 2000.

Clearly there has been a persistent increase in energy efficiency but this has not been consistently steady. The curve used in elasticity calculations to smooth out the year-by-year fluctuations in a time series is an exponential curve, sometimes called a growth curve or geometric progression. This models

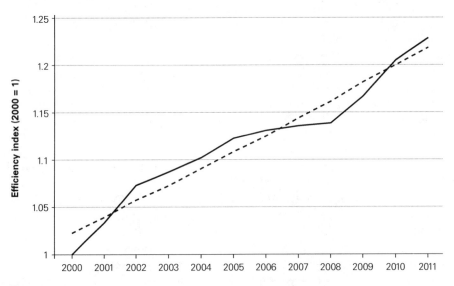

Figure 4.5
Time series of energy efficiency of Portugal's housing stock, 2000–2011, with modelled exponential curve

NOTE: Data from Portugal is used to illustrate how the actual data points for energy efficiency link up. Since values for energy efficiency are relative to any chosen datum, the data has been indexed to the value 1 for the year 2000 (solid line).

the year-by-year increase as if it were proportionately the same each year – as if these homes' energy efficiency increased each year by the same percentage. This curve has the form:

$$V = A \times B^t \tag{4.7}$$

where V is the modelled value for energy efficiency, A is its value in the first year, B is the annual proportionate increase in its value, and t is the number of years since the first year.

The values for these variables are obtained by finding the equation of this type that gives the least total deviation from all the data points along the way – an operation known as a least squares exponential regression. A straightforward procedure for obtaining this regression curve is explained in the Appendix, Section A.7. In this case the curve for Portugal's energy efficiency is:

$$\varepsilon = 1.00743 \times 1.01603^t \tag{4.8}$$

where ε is efficiency and t is the number of years since 2000.

This curve is shown as the dotted line in Figure 4.5.

Turning now to the reduction in energy consumption over the same years, the same operation produces the curves shown in Figure 4.6. Again the continuous line represents the actual values, while the dotted line is the exponential regression model. Here the modelling curve is:

$$E = 0.99631 \times 0.98797^t \tag{4.9}$$

where E is energy consumption.

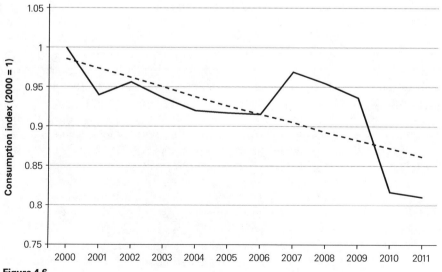

Figure 4.6
Time series of energy consumption of Portugal's housing stock, 2000–2011, with modelled exponential curve

In the Appendix, Section A.3 it is shown that the elasticity rebound effect is obtained from these equations using the formula:

$$R = 1 + \frac{\ln\left(B_{consumption}\right)}{\ln\left(B_{efficiency}\right)} \tag{4.10}$$

For Portugal this is:

$$R = 1 + \frac{\ln\left(0.98797\right)}{\ln\left(1.01603\right)} = 0.239 \; or \; 23.9\% \tag{4.11}$$

The rebound effect for the Portuguese housing stock over the period 2000–2011 was therefore 23.9 per cent. This means that for every 1 per cent increase in the energy efficiency of this stock, 23.9 per cent of this increase went to increasing the quantity of energy services consumed – warmer rooms, more hours watching television, etc., plus whatever portion of this was wasted due to other socio-technical effects. Meanwhile the remaining 76.1 per cent went to reducing the quantity of energy consumed.

These figures are important for national energy planning. The steady energy efficiency improvements in Portugal's housing stock have not led to corresponding, one-for-one reductions in energy consumption. Around 24 per cent of the energy efficiency improvements have been lost to the rebound effect.

3.3 Rebound effect results
The rebound effect results for all 29 countries are displayed in Figure 4.7.

Some important issues arise from these results. To begin with, the EU as a whole shows a rebound effect of 4.8 per cent, which may seem very low, since empirical studies of domestic heating and appliance rebound effects in large western European countries generally give much higher results, as illustrated in Chapters 3 and 4. However, as was noted above, the reduction in energy consumption in 2000–2011 was most likely not all due to energy efficiency upgrades. Possibly around half of it was 'free', arising out of more frugal heating behaviour. Most of this was probably caused by demographic changes, and some may be a response to environmental awareness, since large EU countries such as Germany and the UK have benefited from environmental consciousness-raising over the last decade. If energy consumption reductions as a result of efficiency increases represented half the total energy consumption reductions and these were not counted in the time series data, it can be shown that the rebound effect for the EU as a whole would rise to 53 per cent.

Similarly for Germany, with a rebound effect result here of 10.76 per cent, this would rise to 55 per cent if, as suggested above, up to half the energy consumption reductions are caused by factors other than energy efficiency improvements. This is a plausible value because there is strong evidence that

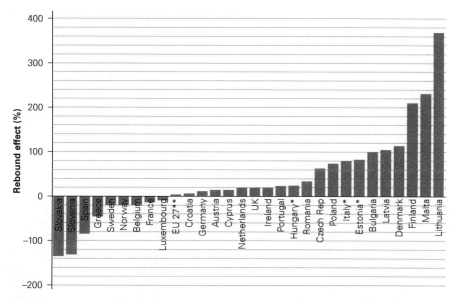

Figure 4.7
Rebound effect for domestic energy services consumption, EU 28 and Norway, 2000–2011
Data source: Odyssee, 2012, author's calculations

the rebound effect for domestic heating alone in Germany is around 36 per cent (see Section 4 below), while it is widely accepted that energy consumption from electrical appliances in German homes is steadily increasing. This increase would boost the rebound effect, as a combined effect of domestic heating and electricity consumption, to something over 40 per cent.

Second, the new EU lands show a much wider spread of rebound effects than the older and more economically developed EU lands. The two countries with the highest rebound effects, Lithuania (370 per cent) and Malta (231 per cent), and the two with the lowest, Slovakia (−135 per cent) and Slovenia (−131 per cent) are new EU countries. Most of the more developed EU lands have rebound effects between −20 per cent and 25 per cent, though rebound effects for Spain and Greece are very low, at −86 per cent and −46 per cent respectively, while those for Finland and Denmark are very high, at 211 per cent and 114 per cent. It is possible that the high results for some new EU lands could be due to households recently becoming wealthy enough to heat their homes comfortably and buy and use a range of electrical appliances. The surprisingly low rebound effects for Spain and Greece could be due to a new emergence of poverty, in which households cannot afford to heat their homes and use electrical equipment to the extent they have done in the past.

Third, the results offer a reminder that wherever there are rebound effects higher than zero, policymakers will miss their energy consumption and CO_2 emission reduction targets if they calculate these on the basis of energy efficiency improvements alone. Instead they need to factor in the rebound effect. The energy efficiency increase they will need to aim for in order to meet a particular consumption reduction goal can be calculated using the formula (derived in the Appendix, Section A.11, equation [A40]):

$$Q = \sqrt[R-1]{1-N} - 1 \tag{4.12}$$

where Q is the actual energy efficiency increase required to meet the policy goal; R is the rebound effect; and N is the energy reduction that is aimed for (all parameters are expressed in decimal form rather than as percentages). For example, in the UK the rebound effect is just under 20 per cent. If UK policymakers are aiming for an 80 per cent reduction in energy consumption, they must aim to increase the energy efficiency of the housing stock by:

$$Q = \sqrt[0.2-1]{1-0.8} - 1 = 6.48 \; or \; 648\% \tag{4.13}$$

Using equation (A42) in the Appendix shows that this is equivalent to reducing the average calculated consumption of the building stock to 13.4 per cent of its current level, rather than 20 per cent. If current average consumption is around 200kWh/m²a including heating and all electrical appliances, this would require a reduction to just under 27kWh/m²a, which is significantly below the level of a passive house for all energy services combined. If there were no rebound effect, the required energy efficiency gain would be 400 per cent, requiring a reduction in consumption to 40kWh/m²a – a more manageable but still very challenging target. Table 4.2 lists all 29 countries, with their year by year changes in energy efficiency and consumption, their rebound effects, and the energy efficiency gains that would be required to reduce consumption by 20 per cent, 30 per cent and 80 per cent in each land. Figure 4.8 shows how the required energy efficiency increase varies with the rebound effect, to achieve these reductions. The table will be of interest to readers who wish to reproduce the calculations or do their own further modelling. As the table shows, equation (4.12) does not work for rebound effects of 100 per cent or more, because these represent a situation where any increase in efficiency leads to an increase in consumption. Also, for rebound effects above about 20 per cent, achieving consumption reductions of 80 per cent or more becomes unrealistic.

The time series method of calculating the rebound effect has disadvantages due to the inclusion of electricity consumption in national statistics, and the compounding effect of heating consumption reductions due to factors other than efficiency gains. However, the composite picture this gives can offer a realistic estimate of the efficiency gains that will be needed, to reduce energy consumption to the levels aimed for in policies.

4. Cross-sectional methods of estimating the rebound effect
The cross-sectional method of estimating the rebound effect was developed via work on the prebound effect (Sunikka-Blank and Galvin, 2012) and further mathematical modelling (Galvin, 2014a). The mathematics for the cross-sectional method are given in the Appendix, Section A.4. In this method, the actual and calculated consumption of a large, preferably representative sample of homes are plotted on a graph. The power curve that best fits these points is then derived.

Table 4.2 Annual energy efficiency gains and consumption changes in EU 28 and Norwegian housing stocks, 2000–2011, showing rebound effects and energy efficiency increases required to bring consumption reductions of 20 per cent, 30 per cent and 80 per cent.

	Average annual increase		Rebound effect	Rebound effect (%)	Efficiency increase (%) required to reduce consumption by		
	Efficiency	Consumption			20%	30%	80%
Slovakia	1.010	0.977	−1.347	−134.7	10	16	99
Slovenia	1.016	0.964	−1.308	−130.8	10	17	101
Spain	1.006	0.988	−0.857	−85.7	13	21	138
Greece	1.004	0.994	−0.463	−46.3	16	28	200
Sweden	1.014	0.983	−0.196	−19.6	21	35	284
Norway	1.007	0.992	−0.194	−19.4	21	35	285
Belgium	1.020	0.977	−0.192	−19.2	21	35	286
France	1.017	0.981	−0.133	−13.3	22	37	314
Luxembourg	1.018	0.980	−0.099	−9.9	23	38	333
EU 27**	1.015	0.986	0.048	4.8	26	45	442
Croatia	1.008	0.992	0.060	6.0	27	46	454
Germany	1.016	0.986	.108	10.8	28	49	507
Austria	1.013	0.989	0.143	14.3	30	52	554
Cyprus	1.023	0.981	0.159	15.9	30	53	577
Netherlands	1.019	0.985	0.193	19.3	32	56	634

Table 4.2 Continued

	Average annual increase		Rebound effect	Rebound effect (%)	Efficiency increase (%) required to reduce consumption by		
	Efficiency	Consumption			20%	30%	80%
UK	1.018	0.986	0.196	19.6	32	56	639
Ireland	1.021	0.984	0.197	19.7	32	56	642
Portugal	1.016	0.988	0.239	23.9	34	60	729
Hungary*	1.006	0.995	0.266	26.6	36	63	795
Romania	1.017	0.989	0.339	33.9	40	72	1,042
Czech Rep.	1.018	0.994	0.635	63.5	84	166	8,105
Poland	1.019	0.995	0.737	73.7	133	287	44,902
Italy*	1.012	0.998	0.798	79.8	202	487	292,916
Estonia*	1.010	0.998	0.832	83.2	276	732	1,421,260
Latvia	1.021	1.001	1.038	103.8	No reduction possible with rebound effects of 100% or more, as any increase in efficiency leads to an increase in consumption		
Denmark	1.011	1.002	1.139	113.9			
Finland	1.006	1.006	2.107	210.7			
Malta	1.007	1.009	2.308	230.8			
Lithuania	1.005	1.014	3.700	370.0			

* Data for 2012 missing; calculations based on 2000–2011.

** The EU 27 figure excludes Croatia, the most recent EU member.

Data source: Odyssee (2013), author's calculations.

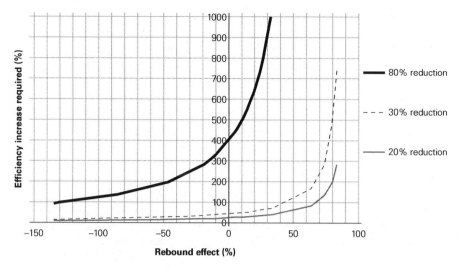

Figure 4.8
Efficiency increase required to reduce energy consumption by 20 per cent, 30 per cent and 80 per cent, for a range of rebound effects

This has the form:

$$E = K \times C^D \tag{4.14}$$

where E is the actual consumption, C is the calculated consumption or EPR, and K and D are constants which are specific to the particular set of data on the graph. The crucial variable here is D, since it can be shown that the average rebound effect for a set of dwellings is equal to $1 - D$ (see Appendix, Section A.4). The way this works is illustrated with the dataset from France, which was briefly discussed in Chapter 3, and reference is also made to the German housing stock, previously discussed in Chapters 1 and 3.

4.1 French housing stock
A plot of actual and calculated consumption in the French housing stock was given in Chapter 3, Figure 3.5. This data comes from a large sample of 913 dwellings which was weighted to take account of discrepancies between the sample and the full French housing stock, in the proportions of different types of buildings and households (Cayla *et al.*, 2010). Therefore it is highly likely that the points on the graph are a fair representation of actual and calculated consumption throughout the entire housing stock. The equation of the power curve that best fits this data is:

$$E = 8.2058C^{0.5013} \tag{4.15}$$

Here the average rebound effect is just under 0.499 or 49.9 per cent, as $D =$ 0.5013. This indicates that for every 1 per cent increase in the energy efficiency of the French housing stock, about 49.9 per cent of this increase is used to provide more energy services – warmer rooms, longer heating periods, etc. – and only the remaining 50.1 per cent is used to reduce energy consumption.

The claim that the rebound effect can be calculated from this data may seem at first to be a contradiction of statements in Section 2 of this chapter, that four data points are needed for a dwelling in order to calculate its rebound effect: the actual and calculated consumption before and after an energy efficiency upgrade. Here there are only two data points per dwelling: the actual and calculated consumption as these stood at the time the dwelling was observed. Further, there is no indication as to which dwellings have had energy efficiency upgrades and which have not. The assumption here, however, is that whenever a large number of dwellings with any particular average calculated consumption level are thermally upgraded, their average actual and calculated consumptions will move leftward along the best-fit curve on the graph. This assumption is justified by the evidence of the actual data points, which can be thought of as a snapshot of energy efficiency and consumption at the particular point in time when the data was gathered. A certain portion of dwellings are continually being thermally upgraded to a greater or lesser extent, and the average of those that had just been upgraded at the time the snapshot was taken is likely to lie on the best-fit curve, unless there is some unknown reason why these households should behave very differently, on average, from others in similar dwellings.

Hence an 'average' French dwelling can be conceived as being upgraded from any point on the right side of the best-fit curve, to any point toward the left side of the curve. This gives the four points for the calculation of this dwelling's rebound effect, as outlined in Section 2 of this chapter. One does not need to derive rebound effect equations individually for different start and end points along the line of the graph, however, as they are all represented by equation (4.15) above.

The weakness of this cross-sectional approach is that it is a snapshot of one moment in time, which does not account for large scale changes over time. What if, for example, French householders are reducing their heating energy consumption not only due to retrofits, but also due to steady improvements in household management, environmental awareness, ever-tightening budgets, or demographic changes? These influences do not show up in the cross-sectional method, leading to possible discrepancies between the two sets of results (see discussion in Section 4.2).

4.2 German housing stock
In Chapter 3, Section 2 it was shown that the power curve which most closely approximates the plot of average actual and calculated consumption for a large, comprehensive dataset of the German housing stock is:

$$E = 4.5C^{0.641} \qquad\qquad (4.16)$$

Using the method above, this indicates a rebound effect of $1 - 0.641 = 0.359$, or 35.9 per cent. Hence for every 1 per cent increase in energy efficiency throughout the German housing stock, 35.9 per cent of this increase is used to provide more energy services while the remaining 64.1 per cent is used to reduce energy consumption.

It is interesting to compare the results for Germany and France obtained by this cross-sectional method, with those obtained by the time series method. The cross-sectional method includes only heating energy efficiency and consumption, while the time series method includes the efficiency and consumption of electrical goods. The inclusion of electrical goods should lead to slightly higher rebound effect results with the time series method, as consumption through electrical goods is increasing throughout Europe. However, because of the time lag in this method, the long-term effect of households heating more frugally over time (Galvin and Sunikka-Blank, 2014b) is tending to reduce heating energy consumption by about as much as the effect of efficiency upgrades. The overall effect is that rebound effects with the time series method would be significantly lower than with the cross-sectional method. Table 4.3 shows rebound effects for Germany and France obtained by these two methods.

The discrepancy between the two results for Germany is easily explained by the empirical finding that over the period 2000–2011 there was approximately twice as much heating energy reduction in German homes as can be explained by energy efficiency improvements (Galvin and Sunikka-Blank, 2014b). If the time series energy consumption reductions for Germany are halved, to give an average annual rate of change of 0.9930, this gives a time series rebound effect for Germany of just under 56 per cent. Since the increasing level of electrical appliance consumption is included in this, the result now seems reasonably consistent with that of 35.9 per cent for the rebound effect due only to heating.

However, the large discrepancies between the results for the French housing stock lead to doubt as to whether the two methods are in fact mutually compatible. It seems unlikely that the cross-sectional data is at fault, as this was gathered and verified by a research team with a high degree of transparency and a number of cross-checks. For the time series data, energy consumption figures are likely to be accurate because this is easy to measure. However, the energy efficiency of domestic heating systems is difficult to estimate on a national scale, and it is possible that the gains in this area have been underestimated in Odyssee data. Rebound effect results in the time series method are highly sensitive to small differences in these figures, so that high accuracy is required if such results are to be reliable.

Table 4.3 Comparison of rebound effect results for German and French housing stocks using time series and cross-sectional methods.

	Rebound effect (%): time series method	Rebound effect (%): cross-sectional method
Germany	10.8	35.9
France	−13.3	49.9

5. Proxy methods: price elasticity and the rebound effect

In Chapter 2 it was suggested that a major cause of the rebound effect is households' response to the effective reduction in the *price* of home heating as a result of an energy efficiency upgrade. Although the 'exogenous' price of energy (the price per unit of *energy* consumed) does not change as a result of an upgrade, the 'endogenous' price (the price per unit of energy *services* consumed) becomes lower. Sorrell and Dimitropoulos (2008) illustrate how this can be used to estimate rebound effects even if no upgrade takes place. If the *exogenous* price of heating fuel changes, the response of households can give an indication of what the rebound effect would be if the *endogenous* price were to change as a result of an energy efficiency upgrade.

This method is fraught with difficulties and pitfalls. The rebound effect in domestic heating is not merely a response to cheaper energy services. Other influences include the dwelling's new and unfamiliar thermal characteristics (often it just seems to heat itself, without any user intervention); difficulties in operating the new heating controls; mismatches between household rhythms and the heating up or cooling down times of the new heating system; technical mismatches between different aspects of the heating technology; and miscalculations of the technology's required size and energy demand (see Chapter 2).

One of the weaknesses in academic discussion of the rebound effect is that it is most frequently conceived, modelled and explained through its effects in private motor vehicle usage (e.g. Berkhout *et al.*, 2000; Greening *et al.*, 2000; Maxwell and McAndrew, 2011; Sorrell, 2007; Sorrell and Dimitropoulos, 2008). When a driver buys a newer, more energy efficient car, the price of driving each kilometre reduces, so in response the driver often drives more kilometres. This is the rebound effect, pure and simple.[1] If, on the other hand, the driver keeps her old car but the exogenous price of fuel falls, she might also drive more kilometres, again in response to a lower price per kilometre. This type of symmetry offers a useful way of predicting the rebound effect when change has taken place in the exogenous price of fuel, rather than the efficiency of the car. The only assumption that is required is that the driver's behavioural responses to exogenous and endogenous fuel prices are the same.

Unlike newly retrofitted homes, however, cars do not drive themselves without user intervention; the basic controls in a new car are the same as in an old car; there are seldom mismatches between user needs and car performance; and the various technical components of a car are usually optimally designed to work well together. Therefore using fuel price response as a proxy for the rebound effect may work well in private motoring, but it does not follow that it works well in domestic heating. At best, it might give a prediction of the component of the rebound effect that is caused by the reduction in the endogenous price of fuel, but it leaves other effects untouched.

The basic concept used in such studies is 'fuel price elasticity'. This is defined as the ratio between the proportionate change in fuel price, and the proportionate change in energy services consumption (or in some definitions, energy consumption). If, for example, each 1 per cent reduction in the price of petrol leads to a 0.1 per cent increase in kilometres driven, the fuel price elasticity of petrol energy services is −0.1, or −10 per cent. This metric is widely used in economics, and results are quite easy to obtain. All that is required is to track and record consumers' mileages as prices change. Further, since mileage is more or less proportional to the quantity of fuel consumed, the same result can be obtained just by tracking petrol sales volumes against price changes.

This can be used as a proxy for the rebound effect if the proportionate change in fuel price is seen as a mirror image of a proportionate change in energy efficiency. An energy efficiency increase of 10 per cent leads to a 10 per cent reduction in the endogenous price of fuel, and it is assumed that this has the same effect on driving behaviour as a 10 per cent reduction in the exogenous price of fuel. Hence the rebound effect R is simply the numerical negative of the fuel price elasticity ξ:

$$R = -\xi \tag{4.17}$$

When this is applied to domestic heating, the assumption is that a certain percentage decrease in the price of heating fuel has the same behavioural effect as the same percentage increase in energy efficiency. It is assumed that in both situations energy services will increase by the same proportion. Further, since energy services consumption is more or less proportional to actual energy consumption (see Chapter 1), the price elasticity of *energy* consumption can be used to estimate the rebound effect, even though this is an elasticity of energy *services* consumption.

One of the most comprehensive examples of such studies is Madlener and Hauertmann's (2011) investigation of rebound effects in the German housing stock. These authors used data from the German Socio-Economic Panel (SOEP), a longitudinal research project conducted by the German Institute for Economic Research (*Deutsches Institut für Wirtschaftsforschung*: see www.diw.de). This sampled the same 11,000 households every year from 1994 to 2012, enabling changes in heating fuel consumption to be tracked alongside changes in heating fuel price. As well as personally interviewing each household member, the survey includes a questionnaire, with questions about heating costs, the size and nature of the dwelling, and household composition. A unique feature of Madlener and Hauertmann's study was that it distinguished between rebound effects in rental and owner-occupied properties, and among households in different income groups. The authors also adjusted heating consumption according to climate in the year prior to the survey, so as to minimise the effects on consumption of year to year outdoor temperature variations.

Average rebound effects were found to be 12 per cent for owner-occupied homes and 40 per cent for rental homes. Income level made almost no difference for owner-occupied homes, but in rental homes the rebound effect ranged from 31 per cent for high income tenants to 49 per cent for low income tenants. The authors suggest that one reason rebound effects are higher for low income households is that they tend to under-heat when fuel prices are high, with the effect that their increase in energy consumption becomes proportionately greater when fuel prices fall.

In Germany 53.3 per cent of homes are owner-occupied while the remaining 46.7 per cent are rental. Weighting the average rebound effect results for owner-occupied and rental homes by the factors 0.533 and 0.467, respectively, gives an average Germany-wide rebound effect of 25.1 per cent. This accords reasonably well with the value of 35.9 per cent found by the cross-sectional method, because it only includes price effects, and not the other rebound effect determinants listed in Chapter 2. The cross-sectional method provides no way of disaggregating the influences of the various determinants of the rebound effect. It simply indicates the total proportion of an energy efficiency upgrade that is 'lost', i.e. that does not go toward reductions in energy consumption.

Although these price elasticity results are consonant with other results, there is still a basic problem with using price elasticity as a proxy for the rebound effect in domestic heating. It assumes households are in fact reacting to (endogenous) fuel price changes when they increase their energy services after a retrofit. This is by no means proven. This author's interviews with a large number of householders in the UK and Germany suggest other factors dominate heating behaviour after a major thermal upgrade. The dominant factor is usually that the home is now a very different thermal environment from before. It is altogether warmer, almost regardless of how the occupants control the heating system. This is not like driving a car, where control is directly achieved through pressure on the accelerator.

6. Conclusions

The four main methods were presented for estimating the rebound effect in domestic heating. The definition of the rebound effect as an elasticity works effectively for all four, and because of its mathematical robustness the results for each method can be coherently compared with one another.

The *case study method* showed how (elasticity) rebound effects can be calculated for specific buildings. This method can be used very effectively to compare rebound effects in identical dwellings with identical retrofits. This can help separate the user and socio-technical effects from purely technical effects, such as poor insulation or mismatches between different pieces of equipment. There is a need for more case study results, as this would provide more certainty as to the range and distribution of rebound effects in the housing stock, and also enable cross-checks with results from the cross-sectional method.

In a case study method the rebound effect as an *EPG* can be found when there are only two known parameters, the actual and calculated consumption after the energy efficiency upgrade. The limitations of the EPG are that it provides no information about behavioural change as a consequence of an energy efficiency upgrade, and its results can be widely different from rebound effects calculated using other definitions. However, as a supplement to the elasticity rebound effect the EPG can be very useful, as it gives a different picture of how energy savings are being played out.

When a further parameter, actual consumption prior to an upgrade, is known in a case study, the rebound effect as an *ESD* can be calculated. This provides very limited information as to behavioural change, and can give nonsensical results if the parameters are not in a particular order in relation to each other. When the fourth parameter, the pre-upgrade calculated consumption, is known, the *elasticity rebound effect* can be calculated. This gives coherent results in all cases, regardless of the values of the parameters. Its results can be coherently compared to elasticity rebound effect results obtained by other methods and in other spheres, and it takes fully into account the effects of occupants' behavioural change as a consequence of an energy efficiency upgrade.

The *time series method* also uses the 'elasticity' definition of the rebound effect. It enables rebound effects to be estimated from large datasets of housing stocks, e.g. on a national scale, when data for average energy efficiency and consumption are available over a number of years. A weakness of this method is that the available datasets combine domestic heating and electrical appliance consumption, and this needs to be remedied. Another weakness is that consumption changes due to factors other than energy efficiency changes are inherently infused in the data. A possible further weakness is that it is very difficult to get accurate data for nationwide increases in the energy efficiency of homes. Where this accuracy is assured, the results this method produces are very useful for policymakers because a simple mathematical transformation can reveal how great the improvements in energy efficiency will need to be, for specific consumption reduction goals to be achieved.

The *cross-sectional method* also uses the elasticity rebound effect. It provides a direct route to estimating the average rebound effect for a large dataset of dwellings when actual and calculated heating consumption are known for each dwelling. Since it uses only these parameters and represents a snapshot in time, its results pertain entirely to heating consumption and are not affected by behavioural changes over time. For these reasons it appears to be the most reliable method for finding the *average* rebound effect for large datasets *at a particular point in time*. Large datasets of actual and calculated consumption are needed for the cross-sectional method, and these are gradually becoming available as more empirical studies are undertaken.

The *proxy method* uses data on price elasticity of heating energy consumption to estimate rebound effects, again using the elasticity definition. Its main usefulness is that it provides a way of predicting rebound effects where there is no information on changes in energy efficiency, or where energy efficiency has not changed. One of its weaknesses is its assumption that households' response to a change in the exogenous price of heating fuel is the same as their response would be to an equivalent change in the endogenous fuel price as a result of an energy efficiency upgrade. Its other weakness is that it can only account for rebound effects due to (endogenous) fuel price changes, and not for those due to additional factors such as occupant interaction with a new thermal environment and technology.

This discussion raises important points for policymakers. There need to be more initiatives to support research which measures and records the actual and calculated consumption of large numbers of dwellings. The data from such research forms the raw material for the cross-sectional method of estimating rebound effects. This is a very effective method for finding average rebound effects for energy efficiency upgrades of all magnitudes, for buildings across the full range of calculated consumption ratings. The results it provides are extremely useful to policymakers, as they can be used to calculate how high the average energy efficiency increase needs to be for the housing stock, in order to achieve the energy consumption reductions policymakers are aiming for.

The time series method can offer a further dimension to such a calculation, as it takes into account energy consumption reductions (or increases) that are not the result of efficiency upgrades but may nevertheless be occurring. On a national scale, year-by-year composite heating energy efficiency and consumption figures need to be collected, separately from figures for appliances, so that robust calculations of rebound effects in domestic heating can be made from this data.

Strong support is also needed for research at the case study level. These studies go beyond mere averages and indicate the full range of magnitudes of rebound effects. Researchers also need to be encouraged to investigate both the magnitude of the rebound effect, and the behaviours of occupants which may be causing or contributing to it. This will improve the understanding of micro-level interventions which could reduce rebound effects and thereby bring energy savings closer to their theoretical potential.

Note

1   Sorrell (2007) suggests this should be extended to include increases in the tonnage carried by private cars when the fuel price falls. Another conceivable effect of fuel price reductions is increased acceleration in driving, as it is well known that gentle acceleration saves fuel. To this author's knowledge this factor has never been included in a rebound effect study on car travel.

5

REBOUND EFFECTS IN LOW ENERGY DWELLINGS AND PASSIVE HOUSES

1. Introduction

This chapter considers how the energy efficiency and consumption of low energy and passive houses relate to the rebound effect.

To achieve policy goals of deep reductions in domestic heating energy consumption and its associated CO_2 emissions, most homes will need to be retrofitted to high energy efficiency standards. There is already some success in achieving this, though the numbers of deep retrofits are as yet small compared to the size of building stocks (Touminen et al., 2012). Most studies of the rebound effect are concerned with the results of energy efficiency upgrades, small or large, in dwellings that are currently thermally inefficient. But it is also important to ask what can be learned about rebound effects from data to hand about consumption in low energy dwellings.

A unique development in the technology of low energy buildings is the advent of the 'passive house'. A passive house is a specific standard of building developed in the 1980s by Wolfgang Feist, a German building physicist, together with his Swedish colleague Bo Andersen. A passive house differs from a conventional dwelling in that it does not have a conventional heating system. Instead, thick insulation, triple-glazed windows, high air-tightness and appropriate orientation to the sun are calculated to enable most of its heating to come from recycling occupants' body heat and incidental heat from electrical appliances. This is supplemented by a small heat pump inserted in a heat-recovery ventilation system. The calculated space heating consumption of a passive house is 15kWh/m^{2}a, which is less than one-third the current average minimum standard for space heating in Germany, a country which has some of the most stringent thermal building standards in Europe. Although the first passive houses were in Germany, the practice has spread to other parts of Europe and beyond, and there is now a passive house institute in at least one US state (www.passivehouse.ca/passive-solar-design/).

2. Rebound effects in low energy retrofits

A modest sized dataset of retrofits in the UK, most of which achieved low energy standard, is publicly available and useful for a study such as this. The dataset is managed by the Technology Strategy Board (www.innovateuk.org/) in conjunction with a number of other organisations, and is part of the innovation and information project Retrofit for the Future. Of the hundreds of retrofits noted in the database, 53 give their post-retrofit actual and calculated consumption, while 27 of these also give pre-retrofit actual consumption. None give pre-retrofit calculated consumption. Nevertheless, it is possible to explore rebound effects using all three definitions: the energy performance gap (EPG); the energy savings deficit (ESD); and the elasticity rebound effect.

2.1 EPG and ESD compared

Table 5.1 gives the data for the 53 dwellings, together with calculations of the EPG and ESD. The definitions of these metrics are given in Chapter 4. It will be recalled (and see the Appendix, Section A.8) that EPG is given by:

$$EPG = \frac{E_2 - C_2}{C_2} \tag{5.1}$$

where E_2 and C_2 are the post-retrofit actual and calculated consumption respectively. The EPGs for these dwellings are displayed in Figure 5.1. They range from a low of –48.5 per cent to a high of 240.5 per cent, with average 64.0 per cent. Of the 53 dwellings, 35 have EPGs above 30 per cent, 16 above 100 per cent (they are consuming more than twice the calculated amount), and 3 above 200 per cent. This profile is a cause for concern. Although these are not 'elasticity' rebound effect figures (and therefore do not give information about how household consumption behaviour has changed as a result of retrofitting), they do indicate a failure by large margins to reach their targets. This concern is confirmed when the ESD is investigated for the 27 dwellings for which sufficient data is available.

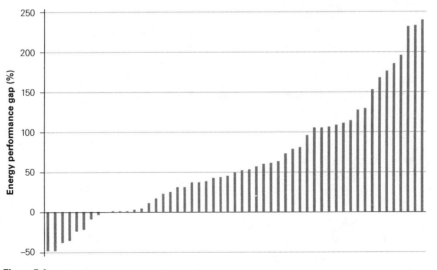

Figure 5.1
Energy performance gaps for 53 UK retrofits
Data source: Retrofit for the Future, author's calculations

Table 5.1 Consumption data from the Retrofit for the Future programme, with author's calculations of EPG and ESD 'rebound effects'.

Dwelling ID	Pre–retrofit measured consumption (kWh/ m2a)	Post– retrofit calculated consumption (kWh/ m²a)	Post– retrofit actual consumption (kWh/ m²a)	Energy per– formance gap (%)	Energy savings deficit
28		46.9	48.8	4.1	
23		28.9	51.6	78.5	
31		28.5	58.6	105.6	
19		114.0	58.9	−48.3	
110		48.1	69.9	45.3	
54		59.6	74.9	25.7	
314		96.4	75.4	−21.8	
72		76.8	76.9	0.1	
10		37.7	78.0	106.9	
50		78.5	79.5	1.3	
51		84.0	83.4	−0.7	
85		66.8	88.0	31.7	Not applicable
33		59.5	97.4	63.7	
32		61.2	98.0	60.1	
22		28.9	98.4	240.5	
52		47.2	101.3	114.6	
86		53.6	113.3	111.4	
93		40.4	115.5	185.9	
8		75.7	122.5	61.8	
73		38.3	127.2	232.1	
108		48.3	129.5	168.1	
96		58.1	160.5	176.2	
25		126.7	173.6	37.0	
24		142.5	175.9	23.4	
27		113.0	195.4	72.9	
93		96.1	221.2	130.2	
32	487.9	144.9	74.6	−48.5	−20.4956
186	315.1	151.5	93.6	−38.2	−35.3912

Table 5.1 Continued

Dwelling ID	Pre–retrofit measured consump–tion (kWh/m2a)	Post–retrofit calculated consump–tion (kWh/m²a)	Post–retrofit actual consump–tion (kWh/m²a)	Energy per–formance gap (%)	Energy savings deficit
42	534.5	119.2	108.8	–8.7	–2.50421
54	550.1	93.2	109.3	17.3	3.5
106	577.5	113.8	109.9	–3.4	–0.8
61	372.2	74.7	114.1	52.7	13.2
315	315.1	151.5	115.4	–23.8	–22.1
81	387.0	91.8	120.3	31.0	9.7
199	284.1	107.8	120.4	11.7	7.1
270	288.7	144.3	146.8	1.7	1.7
77	378.8	97.4	149.2	53.2	18.4
181	287.7	76.3	160.0	109.7	39.6
20	430.0	116.9	160.0	36.9	13.8
278	221.9	118.1	170.6	44.5	50.6
19	321.1	165.3	172.4	4.3	4.6
4	678.4	90.8	186.8	105.7	16.3
127	550.6	75.5	191.3	153.4	24.4
67	263.3	108.8	197.9	81.9	57.7
165	343.0	152.9	211.0	38.0	30.6
136	278.9	148.8	212.8	43.0	49.2
113	701.6	377.0	242.4	–35.7	–41.5
112	395.7	156.5	245.4	56.8	37.2
39	179.6	134.1	263.3	96.3	284.0
12	764.7	79.9	266.1	233.0	27.2
15	541.5	180.6	268.0	48.4	24.2
167	226.9	118.0	269.1	128.1	138.8
131	284.2	95.8	283.8	196.2	99.8
mean	405.9	98.9	142.2	64.0	30.7
std dev	157.5	55.1	63.1	75.1	63.1

The ESD (see Appendix, Section A.9) is given by:

$$ESD = \frac{E_2 - C_2}{E_1 - C_2} \qquad\qquad (5.2)$$

where E_1 is the pre-retrofit actual consumption. As the EPG and ESD are defined and calculated in different ways, the results can be expected to diverge. Figure 5.2 displays the EPG and ESD for the 27 dwellings for which all three of the necessary parameters were given. The results are shown in ascending order of the dwellings' post-retrofit actual consumption.

The ESDs range from −41.5 per cent to 284.0 per cent. The average is 30.7 per cent, far lower than for the EPG. This is deceptive, however, as most of the dwellings for which sufficient information was available to calculate the ESD had low EPGs: the sample of ESDs is biased toward the low end.

A further issue is the divergence of results between EPG and ESD. The EPG is much higher than the ESD for dwellings with IDs 54, 81, 61, 4, 127, 12, 181 and 131, and much lower for dwellings 32 and 39. Divergence is to be expected because the two measures are different metrics based on different parameters. The ESD gives an indication of the *shortfall in energy savings* compared to *what was expected to be achieved* from the starting point of the actual pre-retrofit consumption. On average in this group, just over 69 per cent of expected savings were achieved, while the remainder were lost through over-heating or other rebound effect determinants such as those outlined in Chapter 2. In contrast, the EPG gives the *excess consumption* after a retrofit, in relation to the *level of consumption the retrofit was meant to achieve*. In both cases, rebound effects seem excessive, even though they are measures of different things.

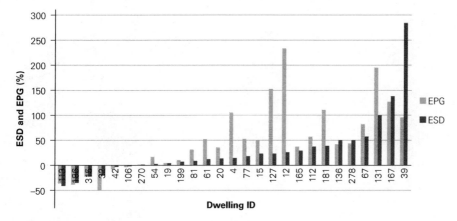

Figure 5.2
Rebound effects as energy performance gap and energy savings deficit (percentage) for 27 UK low energy retrofits

2.2 Consumption goals and the EPG

The EPG can be further investigated to see if there is any relationship between this version of the rebound effect, and the calculated post-retrofit consumption. Does the EPG get larger or smaller as the ambition level of a retrofit increases?

Figure 5.3 gives a plot of EPG against the ambition levels of the retrofits, i.e. the calculated consumption they were aiming to achieve. Dwelling No. 113 was excluded from this dataset as it is clearly an outlier, with a calculated consumption more than twice as high as the next highest. This does not affect the results significantly, however, either in these calculations or the elasticity rebound calculations in Section 2.3.

The continuous line in Figure 5.3 is the logarithmic curve of best fit, while the heavy dotted line that is almost co-linear with the continuous line is an exponential curve closely fitted to this. These curves give the approximate, average EPG for any particular ambition level (i.e. calculated consumption) in the range 25–190kWh/m²a.

The curves indicate that as the ambition level increases (i.e. as lower consumption is aimed for), the EPG rises. For a modest ambition level of about 190kWh/m²a the EPG is zero, i.e. the occupants tend to consume the level calculated. For a more ambitious project aiming for 100kWh/m²a, the EPG is 50 per cent, i.e. households tend to consume half that much again. For a very ambitious project aiming for 28kWh/m²a, the EPG increases to 150 per cent, i.e. consumption is 2.5 times the level calculated. These are averages, of course: not every household consumes that much, and some consume a lot more.

A concern from this analysis is that it is not just that EPGs are high among these retrofits, but that they rise as the ambition level increases. In Chapter 4, Section 3.3 there was discussion of policymakers' goals for domestic energy consumption. It was seen that the rebound effect (as an elasticity) makes it harder for these goals to be reached. Similar calculations are possible with the EPG. If, for example, policy aims to achieve average actual consumption of 60kWh/m²a in the UK building stock, these results suggest the average calculated consumption would have to be 25kWh/m²a. By transforming and

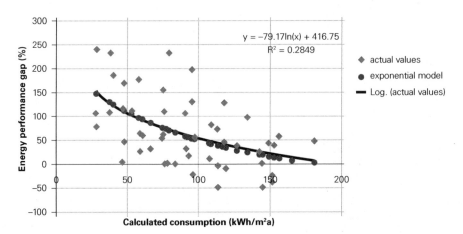

Figure 5.3
Energy performance gap, 52 UK dwellings, with log model (given) and exponential model
243.9*0.9888^ p – 30

then inverting the exponential equation derived for Figure 5.3 it is possible to plot a function[1] that indicates the average calculated consumption levels which engineers would have to aim for, in order to achieve policy goals of various average actual consumption levels. This function is shown in Figure 5.4.

For example, to achieve an average actual consumption of 80kWh/m²a (the horizontal axis in Figure 5.4), engineers would need to plan retrofits to have an average calculated consumption of 35kWh/m²a (the vertical axis). This would be an extremely ambitious goal for the UK housing stock. To achieve actual consumption levels lower than this would be practically impossible, given what is known about the difficulties incurred in retrofitting old buildings to such standards (see, e.g. Galvin, 2013, 2014a; Touminen *et al.*, 2012).

Of course, the data these calculations are based on are limited to 53 UK retrofits. However, their values, interrelationships and results are credible in comparison with comparable data from other countries.

2.3 Elasticity rebound effect with Retrofit for the Future data

The elasticity rebound effect can be calculated for this dataset using the cross-sectional method outlined in Chapter 4. Since post-retrofit actual and calculated consumption are known for all these dwellings, these are plotted and the best-fit power curve is found. This is displayed in Figure 5.5. The power curve for the 52 data points is:

$$E = 12.813C^{0.5168} \tag{5.3}$$

Using equation (A14) derived in the Appendix, Section A.4, the elasticity rebound effect is therefore $1 - 0.5168 = 0.4832$, or 48.32 per cent.

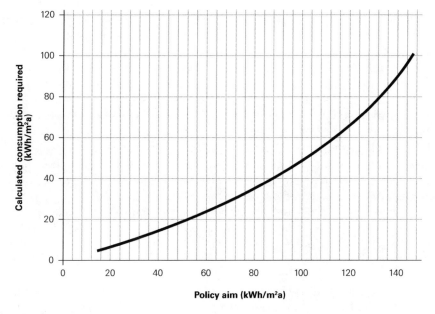

Figure 5.4
Level of calculated consumption required to achieve policy goals of actual consumption levels
Data source: Retrofit for the Future, author's calculations

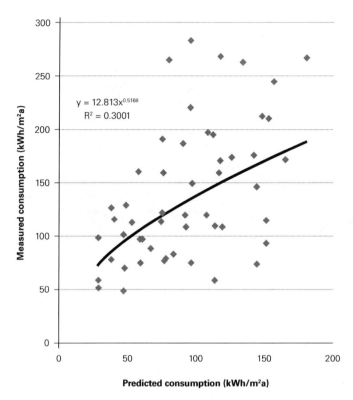

Figure 5.5
Predicted and measured consumption, 52 buildings, showing rebound effect = 1 − 0.5168 = 0.4832 = 48.32%

All the data here are from the post-retrofit situation of these dwellings, so this rebound effect result is not a measure of changes that took place when they were retrofitted. Rather, it suggests the average rebound effect that is likely to ensue if the energy efficiency of a dwelling with a calculated consumption within this range (25–190kWh/m²a) is increased further. The equation for the best-fit curve shows that 48 per cent of this energy upgrade would be used to increase energy services (warmer rooms, more window opening, etc.) while only the remaining 52 per cent would go to reducing energy consumption.

This seems a very high rebound effect, but it is comparable with figures for other countries calculated by the same cross-sectional method. The figure calculated for France's building stock was just under 50 per cent while that for Germany was just under 36 per cent (Chapter 4).

This is a very small dataset which cannot be claimed to be representative of the UK housing stock. However, it is the only such dataset available on low energy UK retrofits at present so it is important to make preliminary estimates such as this. If policymakers are to be well informed in setting goals for very low energy consumption, dwellings with consumption levels in the range of this dataset will need to be improved further. This preliminary investigation suggests rebound effects are likely to be high, so that calculated consumption goals would need to be stringent to bring consumption down significantly. For example, using equation (4.12) in Chapter 4 shows that, in order to halve the average actual consumption of dwellings in this group, their current energy efficiency would need to be increased by 282 per cent, equivalent to their calculated consumption being reduced to 35 per cent of its current value. This would require the range of calculated consumption to be reduced from 25–190kWh/m²a to 9–67kWh/m²a, an extremely challenging if not practically impossible task.

3. Passive houses and the rebound effect

The first prototype passive houses were built in the early 1990s in Darmstadt-Kranichstein (Feist, 1992) and Gross-Umstadt (Knissel and Feist, 1997) in Germany. Passive houses became commercially successful with the building of a block of passive terraced houses in Wiesbaden, Germany in 1997 (Ebel *et al.*, 2003). Since that time, passive houses have been built in several other European countries (detailed information on the spread of passive house technology and its adoption in policy can be found on the English web pages of the original Passive House Institute: www.passiv.de/en/index.php).

A passive house costs approximately 5–15 per cent more to build than a conventional house of the same dimensions (Audenaert *et al.*, 2008; Mahdavi and Doppelbauer, 2010; Schneiders and Hermelink, 2006), and one of the issues prospective homeowners may find important is whether these extra costs will pay back, through fuel savings, as a result of the superior energy efficiency of a passive house. This issue is discussed in detail in Galvin (2014d).

A passive house is designed to consume 15kWh/m²a for space heating. To put this in perspective, the maximum permissible space heating consumption in the design of a new home in Germany is approximately 58kWh/m²a (this varies depending on the dimensions of the building). The average actual space heating consumption in German homes is around 135kWh/m²a (Schröder *et al.*, 2011). Water heating adds a further 12–15kWh/m²a, but this is not considered in this analysis because this figure seems to be constant across the housing stock.

The passive house, then, is a major advance in thermal building technology. Its one major drawback is that its energy source is electricity. To consume 1kWh of electrical energy in a house (called 'final energy') requires about 2.7kWh to be consumed at the power station (called 'primary energy'), due to inefficiencies in generators and losses in transmission lines. For this reason the primary heating energy consumption of a passive house is close to 40kWh/m²a. This can be offset by photovoltaic panels on the roof, but solar radiation is weak and sporadic in [northern European] winters when the heating system needs it. Homeowners can make up some of the costs of winter electricity by using their summer excess photovoltaic power for electrical appliances, but these tend mostly to be run in the early morning and evenings, when solar radiation is sparse. The excess summer electrical energy is then fed into the grid, bringing problems of load mismatches. Hence this analysis, along with all known peer-reviewed studies, makes a clear distinction between the final and primary heating consumption energy of passive houses in analysing their performance.

An analysis of elasticity rebound effects or ESDs is not possible with passive house consumption data. Almost all these dwellings are new, so there are no pre-retrofit consumption figures. Further, all passive houses, by definition, have the same calculated final heating consumption of 15kWh/m²a, so the rebound effect as an EPG can be calculated wherever actual consumption is known. To date at least 14 datasets of passive house performance from four European countries have been critically analysed and the results published in peer-reviewed journals. The details of these and one non-peer-reviewed study are listed in Table 5.2, together with calculations of the average EPG for each dataset. The overall average consumption and EPG are also shown, weighted according to the number of dwellings in each dataset.

The average final space heating consumption for these 229 passive house dwellings is 21.7kWh/m²a, indicating an average EPG of 44.6 per cent. This implies that the average *primary* energy consumption for space heating in these passive houses is approximately 59kWh/m²a. At first sight this may seem to indicate something of a technical failure, since there is a large gap between 15 and 59kWh/m²a.

However, as was shown in Section 2 of this chapter, to achieve an actual consumption level of 59kWh/m²a in a conventional low energy house can require a calculated consumption of 21kWh/m²a. It is much more difficult to achieve a calculated consumption as low as this with a conventional house design, than to achieve a calculated final energy consumption of 15kWh/m²a with a passive house design. Further, the EPG of 44.6 per cent is low compared to EPGs for conventional low energy houses. As Figure 5.3 indicates, average EPGs can range from 50 to 150 per cent for conventional houses which have calculated consumption in the range 25 to 100kWh/m²a. In these respects a passive house is clearly the more successful option for energy saving.

Two further points need to be considered. First, there seems to be an inverse relationship between electricity price and EPG in these passive houses. Table 5.3 gives composite average values for consumption and EPG in the passive house datasets according to their country, together with the cost per kWh of household electricity in each country. There are very few data points here so it must be emphasised that this finding is only tentative. In Germany, where electricity costs are significantly higher than in the other four countries, consumption and the EPG are significantly lower than those in the other countries. The relationship is not linear or smooth, however, since Austria has a relatively high electricity price but also the highest consumption and EPG, and there are not enough countries in the dataset to be confident about a direct relationship or causality between price and consumption. Nevertheless, this suggests the possibility that electricity price could be a factor in Germany's consistently low passive house consumption compared to that in other European countries. Cross-country studies focusing on this issue would be needed to explore this possibility further.

Second, it is worth noting the large spread of consumption and EPG values for passive houses, even those within the same housing estates. Table 5.2 lists these (sixth column). Schneiders and Hermelink (2006) give a pictorial representation of the spread of consumption values in their datasets, and one of these (Kassel, Germany) is reproduced here in Figure 5.6.

In Figure 5.6 it is seen that three households are consuming significantly more than the other high consuming households. These three houses' consumption figures are 47, 35.5 and 37kWh/m²a, while the highest of all the others is 25kWh/m²a with average 14.9kWh/m²a. These three together make

Table 5.2 Location and measured heating consumption of passive houses, from research studies. All are peer-reviewed except the last, which comes from the Passive House Institute.

Author(s) of study	Peer reviewed or not	Location (and country) of dwellings	Number of dwellings	Average measured final heating consumption (kWh/m²a)	Range (kWh/m²a)	Total for group	Average EPG for group (%)
	yes	Hannover (DE)	32	16.0	4–33	512	6.7
		Kassel (DE)	23	17.7	0.5–42	407.1	18.0
		Egg (AT)	4	35.4	23–43	141.6	136.0
		Hörbranz (AT)	3	12.3	6–23	36.9	–18.0
		Wolfurt (AT)	10	21.9	7–32	219	46.0
Schneiders and Hermelink (2006)		Dornbirn (AT)	1	40.7	N/A	40.7	171.3
		Gringl (AT)	6	35.1	24–46	210.6	134.0
		Kunchl (AT)	25	23.9	2–49	597.5	59.3
		Horn (AT)	1	27.9	N/A	27.9	86.0
		Steyr (AT)	3	20.0	16–27	60	33.3
		Luzern (CH)	5	22.4	5–23	112	49.3
Molin et al. (2011)	yes	Linköping (SE)	9	21.0	2–25	189	40.0
Blight et al. (2013)	yes	Various (UK)	100	22.5	1.5–47	2250	50.0
Ridley et al. (2013)	yes	London (UK)	1	15.0	N/A	15	0.0
Peper and Feist (2008)	no	Ludwigshafen (DE)	6	24.4	not given	146.4	62.7
		Total	229				
		Weighted average		21.7			44.6

Table 5.3 Average actual consumption in passive houses in five European countries, with EPGs and electricity costs.

	Average actual consumption (kWh/m²a)	Average EPG (%)	Cost of electricity (€c/kWh)
Germany	16.7	11.4	29.71
Austria	25.2	67.8	20.40
UK	22.4	49.5	17.64
Sweden	21.0	40.0	17.56
Switzerland	22.4	49.3	17.36

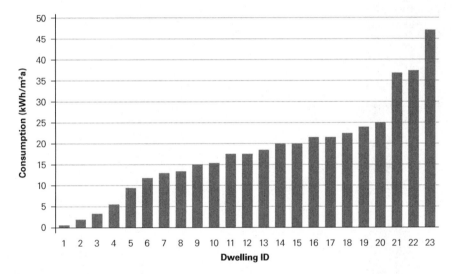

Figure 5.6
Actual heating consumption, 23 passive houses in Kassel, Germany
Data source: Schneiders and Hermelink (2006)

up 10 per cent of the households but are responsible for 30 per cent of the total consumption of the 23 dwellings. This pattern is typical in most datasets of matched low energy dwellings, not only among passive houses but also with conventional houses. In this dataset from Kassel, the ten lowest consuming households, which make up 43 per cent of the homes, consume only 21 per cent of the energy, while 50 per cent of the energy is consumed by the seven highest-consuming households, which make up 30 per cent of the homes.

In a similar vein, the heating consumption levels in three apartment blocks, each consisting of 30 low energy conventional dwellings, were analysed in Galvin (2013) and the same patterns were evident. In each case a small minority of households was responsible for the excessively high average EPG. The study suggested that many of the policy interventions aimed at reducing EPGs are ineffective because they spread their target over all consumers, or all consumers of a particular socio-economic background. It was suggested that a better policy would be to target interventions specifically at households showing high EPGs. In the case of the passive houses in Kassel, for example, inducing just three households to reduce their consumption to the level of the calculated rating would reduce the total consumption of this group of households by 18 per cent and bring its average consumption down below the calculated passive house standard of 15kWh/m²a. To this author's knowledge such an approach has never been tried. Instead, as Gynther *et al.* (2012) show in a Europe-wide survey, the vast majority of intervention programmes in Europe are aimed at high and low consumers alike, based on selection criteria other than propensity to consume excessively, and are largely ineffective in reducing consumption significantly. There is a need for interventions to be developed which target high consumers individually, regardless of their socio-economic characteristics (Galvin, 2013).

4. Conclusions

This chapter has explored rebound effect issues in relation to low energy houses, including both conventional and passive houses. Using a dataset of retrofit results from 53 dwellings in the UK, the rebound effect as an EPG was calculated for each dwelling, and the ESD for each of the 27 dwellings for which sufficient data was available. The magnitude of the EPGs, which averaged 64 per cent and ranged up to 240.5 per cent, was identified as a cause for concern, as this indicates that ambitions to bring consumption down to levels that are theoretically possible are being frustrated. Of further concern was that the EPG increases by an exponential function as the ambition level increases. To achieve an actual average space heating consumption level of 60kWh/m²a, which is about the level of new build (calculated) standards, would require a calculated consumption level of about 24kWh/m²a, which is extremely difficult to design for in retrofits of older houses.

Similar concerns arose when an average elasticity rebound effect was calculated for this dataset. Here the rebound effect was 48 per cent, indicating that, if further energy efficiency upgrades were applied to houses such as these, some 48 per cent of the efficiency gain would be lost to comfort-taking, human–technical interface problems and technical failures, leaving only 52 per cent of the efficiency increase to make inroads into reducing consumption.

A more optimistic picture emerged with passive houses, despite EPGs averaging around 45 per cent. Even though these dwellings supplement their natural heating with electricity, which effectively increases their actual average primary energy consumption to around 59kWh/m²a, the success story is that this average level of heating consumption is significantly lower than for equivalent conventional houses. To achieve this average through retrofitting conventional dwellings or even building new homes, taking into account their large rebound effects, would require designs that are, practically, very difficult to produce.

A new approach to policy interventions has been suggested, in which excessive consumers are targeted for assistance in bringing down their consumption. Even in passive house datasets, there is a great range of variation in actual consumption levels and EPGs among similar or identical dwellings. Often, the excessive consumption is due to a small minority of households, which is biasing the average consumption figures to worrying high levels. A policy focus on excessive consumption would be an alternative to current, largely ineffective practices of targeting all consumers or cross-sections of households based on their socio-economic characteristics. The challenge for policy is to define and implement social norms and practices for energy services – particularly for new technologies (e.g. passive house).

Note

1   The function is: $E = C + (C \times 234.9 \times 0.9888^{\wedge}C - 30 \, xC)/100$. As there is no simple way of inverting this to make C the subject, it was solved here using a spreadsheet.

6

FUEL POVERTY AND THE REBOUND EFFECT

1. Introduction

Fuel poverty is a topic that is highly relevant to the rebound effect. When a household is lifted out of fuel poverty, the indoor temperature of its dwelling is higher than before. This represents an increase in its consumption of energy *services* – even if its *energy* consumption reduces. This is a classic example of the rebound effect. The rebound effect is often mentioned in studies on fuel poverty, but to date there does not appear to be a specific study that links the two topics and works systematically through the issues that inevitably arise. This chapter is a first attempt to fill this gap.

The term 'fuel poverty' came into housing vocabulary in the early 1990s, largely through the insight and efforts of Brenda Boardman, who was at that time a doctoral student working with an interdisciplinary team at Oxford University. There are perhaps two key insights that characterise fuel poverty. First, it can be clearly distinguished from poverty in general, even though these may overlap (MacKerron, 2012). A household may have adequate funds and resources to fulfil its needs for food, clothing, accommodation and everyday demands, but not be able to afford to heat its home to a healthy, comfortable temperature. Boardman insisted this is a special kind of poverty, which she called 'fuel poverty'. Conversely, there are households who can easily afford to heat their homes, even in frigid or temperate climates, but cannot afford other basic necessities such as nutritious food and good clothing. These households are poor, but not fuel-poor.

If this seems an odd juxtaposition of factors, it makes perfect sense when the second basic insight is added: the key, distinguishing cause of fuel poverty is the technical quality of the dwelling the household lives in (Boardman, 2012; Liddell, 2012). Fuel poverty occurs in dwellings which are thermally poor; they are hard to heat; they may be draughty, poorly insulated, given to dampness, face away from the sun, have inefficient heating systems; or all or a mixture of these. While fuel poverty is exacerbated by rising fuel prices and falling incomes, and some households may fall into fuel poverty for the first time when their fuel costs rise or incomes fall, the distinguishing cause is the technical, thermal quality of the building. In Boardman's words: '... while fuel prices and low incomes are constituent factors, the real cause of fuel poverty is the energy inefficiency of the home' (Boardman, 2012).

Fuel poverty is not just a matter of discomfort; it is a matter of life and death, since cold, damp indoor conditions are bad for human health (Geddes *et al.*, 2011; Liddell and Laurence, 2010). In New Zealand, for example, it was found that a far larger share of winter hospital admissions were of people living in thermally poor homes, than the admissions of those in better buildings (Telfar-Barnard, 2010). A major cause of the large excess of winter deaths in New Zealand is also thought to be the very poor thermal quality of most of that country's homes (Howden-Chapman *et al.*, 2012). Findings are similar for the UK (Boardman, 2010). Depending on the definition used, somewhere between 1.5 and 5 million of the UK's 21 million households live in fuel poverty (Moore, 2012), i.e. between 7 per cent and 24 per cent. For Continental Europe there are few comprehensive studies, though the concept of 'fuel poverty' or 'energy poverty' is now beginning to enter policy and academic discourse in most of these countries, and is the subject of EU Commission reports (e.g. IEE, 2009). It is broadly estimated that fuel poverty affects western European countries comparably to the UK, probably more in the south than the north, while the problem is most likely even more widespread, and more complex, in Eastern Europe and the former Soviet republics (Herrero and Ürge-Vorsatz, 2012).

According to recent studies, most of Europe's fuel-poor households have lower than the median income (Bouzarovski *et al.*, 2012). However, a recent research project in Cambridge, UK illustrates how fuel poverty can affect even those on high incomes (Sunikka-Blank and Galvin, 2014). A couple reported:

> *The house was absolutely freezing cold. I have never been as cold as the first winter we were in this house. The house was incredibly cold, the central heating didn't seem to do much to get the house warmer, certainly not up to a reasonable temperature, and there was condensation and mould... I found that many of my colleagues at work said, 'But of course, my house is that cold [too], I can only afford to have the central heating [on] in 2 rooms.'*

The interviewee's colleagues at work, who could not afford to heat more than two rooms because of the thermal poverty of their dwellings, are well-paid professionals. Like others this author has interviewed, it seems that in a good proportion of UK homes no amount of heating lifts the indoor temperature into the comfortable range. This is very frequently found in homes on the non-sunny[1] side of semi-detached pairs or terraced houses, though not exclusively. For households on low incomes the situation is often far worse, because even a dwelling of moderate thermal quality may be impossible to heat comfortably with the funds such a household can allocate to heating.

Thanks to the tradition of research and policy discussion largely triggered and nurtured by the work of Boardman in the UK and a growing number of others, fuel poverty has been firmly on the UK policy agenda for over a decade and is gradually gaining a hold in Continental Europe. The work of Philippa Howden-Chapman and others in New Zealand have brought it firmly into focus there. In both the UK and New Zealand the mitigation of fuel poverty has been the stated motivating factor in government-led and financed initiatives to improve the thermal standards of the homes of low income and at-risk groups, who almost invariably live in older, thermally inadequate dwellings. Many Continental European countries also have long-running initiatives to mitigate cold in homes, though the phrase 'fuel poverty' is only recently being used there. For example, German and Austrian local authorities are continually upgrading their social housing, as are many of the large cooperative or private housing providers in these countries (Galvin, 2011).

Nevertheless, much of the fuel poverty in Europe is found in the private rental sector. An early UK study painted a grim picture of the lives of elderly tenants in private rental accommodation in London (Smith, 1986). One of its most striking findings was that tenants dared not complain to their landlords about the cold and damp – some of which would have been easily preventable – for fear of eviction. A recent study among tenants of private rental housing in Vienna, Austria gives important insights into how such people cope with fuel poverty (Brunner *et al.*, 2012). Some heat only one room; some dress in outdoor winter clothing in the home, including thermal underwear; some often stay in bed during the day; and some use combinations of these strategies. The researchers also found a 'modesty' in many tenants' lifestyles, in which they spend little on clothes, food, travel, entertainment and recreational goods and services, so as to have something left over for the fuel bill. Some fail in this endeavour and are faced with extra bills for late fuel payment, deepening their cycle of fuel poverty. Anderson *et al.* (2012) found similar strategies and failures in low income households living in cold homes in the UK.

The term 'energy poverty' is often used in a similar context to 'fuel poverty'. Normally this broadens the scope of the issue to include households' inability to get the services they need from electrical equipment, such as stoves and washing machines. As Boardman (2010) points out, often those on low incomes not only have thermally poor homes, but their electrical equipment may be outdated and inefficient. Further, in Germany household electricity prices are around twice the EU average (E-Control, 2013), due to the large subsidies domestic consumers must pay to those who produce renewable electrical energy via photovoltaics. This is pushing more and more German households into *'Energiearmut'* (energy poverty) and appears to be motivating groups of academics to embark on serious research into the problem. The first *Energiearmut* conference was held in Kassel in June 2013. But while the broader issue of energy poverty is a serious problem, this book is primarily concerned with domestic heating, hence its focus in this chapter is more narrowly centred on what is traditionally regarded as 'fuel' poverty, the inability to heat one's home to a healthy level.

The remainder of this chapter explores the questions: Why does the rebound effect become an inevitable issue when attempting to mitigate fuel poverty? What level of rebound effect can be expected when fuel poverty is mitigated? What are the gaps in current research on fuel poverty with respect to the rebound effect, and how might these be addressed?

2. Why the rebound effect happens after fuel poverty mitigation

Figure 6.1 shows how the rebound effect operates when a dwelling undergoes an energy efficiency upgrade or retrofit. Starting at the left hand end of the diagram, the dwelling's energy efficiency increases by a certain amount (first arrow on the left). In theory this should cause a fall in energy consumption of an equivalent amount (second arrow from the left), for example a 40 per cent increase in energy efficiency should cause a corresponding reduction in energy consumption. However, the evidence from metered energy consumption after upgrades almost invariably shows that the actual fall in consumption is much less, say 18 per cent rather than 40 per cent. Of the remaining 22 per cent, a certain amount (say 8 per cent in this diagram) is almost always lost due to technical problems with the retrofit, while the final 14 per cent is 'taken back' and used to increase the comfort levels in the dwelling. As was explained in Chapter 2, the distinction between 'energy lost' and 'energy taken back' is not hard and fast. There are gradations in between, for example where energy is lost because occupants cannot operate the new heating system properly, or because the heating up and cooling down times of features (e.g. underfloor heating) do not mesh with occupants' actual needs (for example, if it takes five hours for underfloor heating to cool down or warm up, and the occupants want to go out for eight hours).

Bearing in mind, then, that this diagram simplifies the reality, a certain portion of the theoretical energy consumption reduction, labelled here 'Energy taken back', goes to increasing energy services, shown by the arrow on the right hand side. The important point for the mitigation of fuel poverty is that this increase needs to be very large. Fuel-poor homes have very low levels of energy services, including cold indoor temperatures, short heating periods, few rooms heated, poor ventilation, excessive damp – even if they are consuming large quantities of heating fuel. They need a large increase in energy services to be lifted out of fuel poverty.

Therefore a substantial rebound effect is *inevitable* and *intended* when mitigating fuel poverty. This does not necessarily mean that energy will *not* be saved when retrofitting fuel-poor homes (but for an exception see Hong *et al.*, 2006), but that *not as much* energy will be saved compared to retrofits of other homes (see, for example, Milne and Boardman, 2000; Boardman, 2010; Ürge-Vorsatz and Herrero, 2012). In reviewing the literature on fuel poverty it is clear that this creates an uneasy tension between the desire for climate change mitigation and the desire to mitigate fuel poverty, an issue discussed further in Section 4 of this chapter. Meanwhile, however, the claim that mitigating fuel

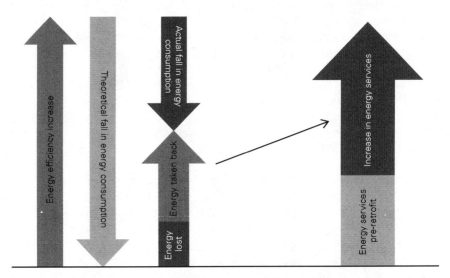

Figure 6.1
Schematic showing the results of the rebound effect when a dwelling undergoes an energy efficiency upgrade

poverty leads to high rebound effects *is not a criticism of the mitigation of fuel poverty.* Rather, research on fuel poverty does not have a comprehensive stream dealing with the rebound effect, and this chapter aims to begin to fill that gap.

3. Fuel poverty and the magnitude of the rebound effect

3.1 High rebounds in upgrades of fuel-poor homes

When a particular dwelling undergoes thermal retrofitting, it is highly likely that there will be a rebound effect, though this might not necessarily be so. However, when a large random sample of dwellings from any housing stock is thermally retrofitted, there is always a significant average rebound effect for that sample (described in Chapters 3 and 4). Any increase in energy services means there is a rebound effect: higher indoor temperatures, longer heating periods, more rooms heated, more generous ventilation, less dampness. Fuel-poor homes are, by definition, low on energy services prior to a retrofit. The aims in mitigating fuel poverty are to bring their energy services up to the level of any good, energy efficient home. *Hence the aim is to give them a large, health-giving rebound effect.*

In a pioneering study at the beginning of the twentieth century, Milne and Boardman (2000) investigated the changes in energy consumption that took place when UK houses with a range of indoor temperatures underwent energy efficiency upgrades. These researchers found that, on average, the colder the house before the upgrade, the higher the proportion of the theoretical energy saving was taken back to increase energy services. If the pre-retrofit temperature was 14°C, about 50 per cent was taken back: the energy saving was only 50 per cent of that calculated. For a pre-retrofit temperature of 16.5°C, about 30 per cent of the calculated savings were taken back. These percentages are not strictly rebound effects, but they give an indication that large rebound effects are manifest. Only where the pre-retrofit indoor temperature was 20°C was the full energy saving realised, an energy take-back of 0 per cent.

Nevertheless, it was not a simple matter of cold homes becoming warmer after a retrofit. In some cases the increase in indoor temperature was as low as 0.5°C since, as the authors pointed out:

> Occupants of cold houses ... may choose to save the energy/money because of financial constraints, or take more benefit as an increase in comfort level if their budgets are not as restricted, leading to wider range of outcomes.
>
> (Milne and Boardman, 2000: 419)

This implies that the energy take-back figures would have been even higher than the above percentages if all the households in cold homes had taken full advantage of their upgrades by heating to 20°C.

This may often be what happens, as the following example illustrates.

3.2 UK Warm Front programme and rebound effects

Hong *et al.* (2006) measured indoor temperature and fuel consumption in over 1,300 houses in five different socio-economic and geographic UK urban areas, in two successive winter periods. These homes were all beneficiaries of the

UK Government's Warm Front programme, which aimed to upgrade the energy efficiency of dwellings of low income and at-risk households – those most likely to be in fuel poverty. Most of these homes underwent energy efficiency improvements in the period between the two winters.

Energy consumption was monitored in both winter periods in 242 of these homes, and energy consumption actually *increased* by 35 per cent. A frequency bar graph for these is shown in Figure 6.2. A significant shift downwards can be observed at the left hand end of the bar graph, which represents low consumption, and upwards at the right hand end, which represents high consumption, for the second year's measurements – i.e. fewer homes consumed less than 50kWh/day after upgrading, and more consumed more than 70kWh/day. This indicates that the increase in consumption was general throughout the sample of homes, and not the result of just a few rogue cases increasing their consumption massively.

A fuel consumption increase after an energy efficiency upgrade would represent a rebound effect of over 100 per cent (see Chapters 1 and 4), and this does not seem to accord with other UK studies such as Milne and Boardman (2000). It can happen, of course, as some dwellings are so thermally poor that occupants hardly bother to heat them, or heat only one room, but then heat more liberally after an upgrade because it now pays to do so (i.e. they get good value for the money they pay for the heating fuel). To investigate this further, Hong and colleagues developed a metric which they called 'normalized space heating fuel consumption' and applied it to approximately 700 of the dwellings. This is the amount of energy required to increase the indoor temperature by 1°C. First, they calculated what this should be for each dwelling before and after their thermal improvements, given the expected effects of these improvements. Their calculations showed that the required amount of energy should decrease, as a result of these houses' thermal upgrades, by an average of around 30 per cent. This seems reasonable, as after a thermal upgrade it should take less energy to increase the indoor temperature by 1°C.

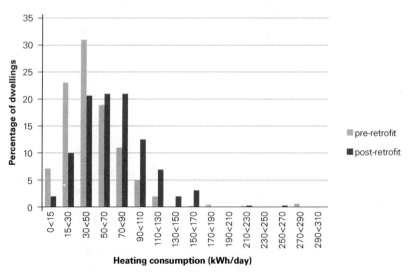

Figure 6.2
Total daily fuel consumption in 242 dwellings, pre-and post-retrofit

When it came to actual measurements in the dwellings, however, they found no significant reduction in the required energy consumption at all. Homes which had required, for example, an extra 40Wh/K/m²/day (120 Watt-hours of energy per square metre of floor area per day) for every 1°C increase in ambient indoor temperature before they were thermally upgraded, required about 40Wh/K/m²/day after their upgrades. This surprising finding is displayed in Figure 6.3.

Again it is clear that the result is not caused by one or two rogue cases, as Figure 6.3 shows the same tendency for every band of required heating energy. In each band there is very little difference between the pre- and post-retrofit amount needed to increase indoor temperature by 1°C. Further, this result cannot be easily explained by household behaviour. Household behaviour changes can easily explain why, for example, a house is 1°C warmer after a thermal upgrade than before, but not why it takes a particular amount of energy to *make* it 1°C warmer.

Hong and colleagues suggested a number of likely reasons for this phenomenon. Cavity wall insulation was often not well installed, as thermal cameras showed large gaps, especially above windows. Homes that switched from electricity to gas suffered air infringement problems as the gas boiler dragged cold air into the dwelling to keep the combustion going. The research measurements did not account for portable electric heaters which may have been used pre-upgrade and then abandoned. It was difficult to disentangle water heating from space heating consumption: water heating is often on a different energy source prior to a major thermal upgrade, so it gets included in post- but not pre-upgrade measurements. Hence it is possible that the methodology for this 'normalised' consumption was simply not good enough for the empirical challenges involved. The researchers might also have missed some important points about post-upgrade household behaviour. Occupants might have opened the windows more often after an upgrade, as found in a research project described in Galvin (2013), thus causing more energy to be consumed in order to raise the temperature by 1°C. They might also have had difficulties interfacing with their new heating technology, a phenomenon this author has seen repeatedly in Continental upgrade situations.

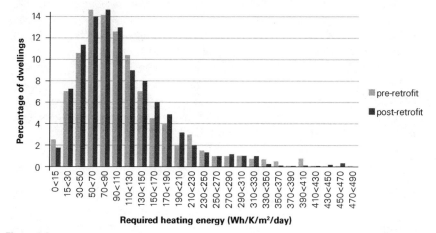

Figure 6.3
Change in 'normalised' actual heating consumption in UK dwellings under Warm Front scheme
Data source: Hong et al., 2006

Nevertheless, it is highly likely that Hong and colleagues were finding similar issues to those outlined in Chapter 2, but to a greater extent where homes were fuel-poor. The deep technical interventions that are required to lift homes out of fuel poverty can be difficult to apply correctly, and do not always mesh well with people's indoor lifestyles. Large rebound effects can be expected in upgrading fuel-poor homes, because of technical failures, user behaviour changes, and the interfaces between these two.

3.3 A general curve for rebound effects in fuel-poor homes

The UK examples above, taken from Warm Front retrofit projects aimed at homes likely to be in fuel poverty, have parallels and differences with those in the Retrofit for the Future project, described in Chapter 5. Retrofit for the Future cases, which included fuel-poor and non-fuel-poor homes, showed energy performance gaps (EPGs) of up to 230 per cent, and over 50 per cent for just under one-third of dwellings. This could suggest that some of the reasons for excess post-retrofit consumption may be similar to those identified by Hong *et al.* (2006) above. There is no direct relationship between pre-retrofit thermal quality and the EPG after retrofitting, but for those buildings where enough data was available to find elasticity rebound effects, the average rebound effect was 62 per cent, which is typical for retrofitting to very low energy standards whether there has been fuel poverty or not.

There is not enough data available to plot a general curve for average rebound effects for UK dwellings, but as was seen in Chapter 3, a very good dataset is available for the French housing stock. Since both the calculated and measured heating consumption are known for all 913 of the dwellings in this dataset, it serves as a useful platform to theorise about rebound effects with fuel poverty.

The data is reproduced here in Figure 6.4, but with two extra curves added. The lower light line represents a possible average of actual heating consumption for each value of calculated consumption, for fuel-poor homes. These homes have low levels of energy services and therefore large prebound effects, i.e. their actual consumption is well below their dwellings' calculated consumption. The aim of mitigating fuel poverty is to reduce this gap, thus increasing their take of energy services. When these homes are thermally upgraded, their calculated/actual consumption levels would be expected to move leftward along the curve, closing this gap more rapidly than the average house, which moves leftward along the continuous curved line. The curve representing these fuel-poor homes has been chosen somewhat arbitrarily, but the more fuel-poor a set of homes is, the lower and flatter their curve would be on the graph – lower because the more fuel-poor a home is the larger its prebound effect is, and flatter because the aim of upgrading these homes is to get them as close as possible to full energy service consumption.

The dotted curve on the graph has the equation:

$$E = 15C^{0.35} \tag{6.1}$$

This implies that upgrading these homes would produce an average elasticity rebound effect of 0.65 or 65 per cent (see the Appendix, Section A.4 and equation [A14]). This is significantly higher than the rebound effect for the average of the whole dataset, which is just under 50 per cent. Fuel-poor homes have higher rebound effects, when upgraded, than average homes.

In contrast, the dotted curve higher in the graph represents a possible case of 'fuel-rich' households, i.e. households who can heat their homes generously. These display relatively low prebound effects prior to retrofitting, and when their homes are thermally upgraded they do not move so sharply toward 100 per cent energy services, as they are already close to this level. This particular curve, again chosen somewhat arbitrarily, has the equation:

$$E = 4C^{0.70} \hspace{4cm} (6.2)$$

This implies an elasticity rebound effect of 30 per cent, much lower than the average rebound effect of just under 50 per cent.

It must be emphasised that this is purely a modelling exercise, with no data on fuel poverty in French homes to substantiate it. However, it seems an intuitively credible approach to the issue, as it takes into account the high prebound effects that are typically found in fuel-poor homes, together with the aim of fuel poverty mitigation which is to reduce these prebound effects drastically. It might be argued that only the data points on the right hand side of the graph should be included in the model, as these represent the most

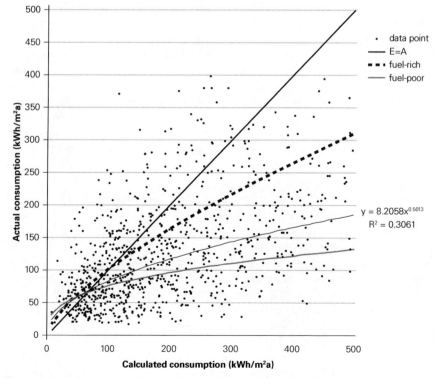

Figure 6.4
French data: estimated space heating consumption as a function of normative calculation, 913 homes, showing average curves for fuel-poor and fuel-rich households
Data source: Cayla et al., 2010, own calculations

technically thermally poor homes. However, both the technical and income dimensions are involved in fuel poverty, and including all the points preserves the interplay of these two determinants of energy affordability. Households succumb to fuel poverty when the technical quality of their dwelling is poor, but if it is only moderately technically poor they may also succumb if their income is too low to get high levels of energy services in their dwellings.

In both modelled and empirical cases, rebound effects after upgrades to mitigate fuel poverty appear to be significantly higher than the norm. This appears to be due in part to technical failures caused by the depth of intervention required to bring a thermally poor home up to a high thermal standard. It is also due to the inherent aim of mitigating fuel poverty: to increase the level of energy services consumption among fuel-poor households. It may also be due to discrepancies in the human–technical interface that result from such drastic interventions. The next section looks critically at how current research on fuel poverty frames the issue of these high rebound effects.

4. Fuel poverty and the rebound effect in current research

Surprisingly, there is very little systematic discussion of rebound effects in upgrades of fuel-poor homes, and this author has not found a peer-reviewed study dedicated specifically to exploring the topic. The central issue that links fuel poverty mitigation and the rebound effect is that, while the elimination of fuel poverty is universally acclaimed as a worthy project, the rebound effect is almost universally perceived negatively. The rebound effects means that less fuel is saved and more CO_2 emitted than expected, thus compromising the goals of energy security and climate change mitigation.[2]

The rebound effect appears in studies of fuel poverty in three main ways. First, some studies simply note that it happens and explore technical or socio-technical explanations for it. Milne and Boardman (2000) and Hong et al. (2006) take this approach, though neither actually calls the phenomenon the 'rebound effect', and it is discussed as something of an afterthought. Milne and Boardman (2000) note that a certain quantity of the energy saved through retrofitting is taken back to increase indoor temperatures, while Hong et al. (2006) attempt to disaggregate the portion taken back through technological failure, from that taken back to increase indoor comfort.

A second response in this stream of research is a kind of stated refusal to engage with the implications of high rebound effects in fuel poverty mitigation. This is exemplified in Ürge-Vorsatz and Herrero's (2012) discussion of synergies between climate change mitigation and energy poverty alleviation. The main thrust of these authors' argument is the very important and convincing insight that reducing fuel poverty and mitigating climate change have significant synergies, since both involve reducing energy consumption. In fuel poverty mitigation there is almost always some reduction in energy consumption, even when the thermally-technically poorest homes are retrofitted (as illustrated theoretically in Figure 6.1), and there certainly is a substantial reduction on average. Therefore, almost any programme to mitigate fuel poverty will save some energy (on average across all the homes upgraded) and help mitigate climate change. These authors also quite rightly note, however, that the high rebound effects in fuel poverty mitigation might appear to compromise this.

Nevertheless, they suggest this should not be labelled a 'rebound effect' (Ürge-Vorsatz and Herrero 2012: 87), since it is inevitable that fuel-poor households will increase their energy services after an energy efficiency upgrade. Therefore, they argue, the rebound effect is not a concept which is applicable to these homes (ibid.: 89).

The logic of this argument can be questioned, since the 'rebound effect' is, by definition, the take-back of saved energy to increase energy services, regardless of whether these were already high or low before the energy efficiency improvement. Accepting these authors' argument would require a major redefinition of the rebound effect. It would seem simpler to leave the definition as it is, and confront the reality of high rebound effects in a constructive way, using a wider contextual framework.

A third approach is offered in Boardman's (2010) wide-ranging and challenging study on fixing fuel poverty. Boardman accepts that high rebound effects can occur when mitigating fuel poverty. In earlier work she had raised the concern that this might deter policymakers from tackling fuel poverty, as this project would not bring high dividends in climate change mitigation (Boardman, 1991). In response to this, Boardman (2010) highlights the important distinction between direct and indirect rebound effects. Direct rebound effects are the increases in energy services that are made in the same area that has been upgraded, such as more warmth in a home that has been thermally retrofitted. Indirect rebound effects are increases in energy services in a second area, such as transport, resulting from money saved when upgrades are performed and energy is saved in a first area, such as a retrofit. Boardman suggests that when fuel-poor homes are upgraded, direct rebound effects often happen but there is much less risk of indirect rebound effects:

> In cold homes, the savings that are being taken as additional warmth reduce the opportunity for a rebound into other energy expenditures. For house-holds who are already warm, the energy efficiency improvement releases money that could be spent on other energy-using activities, such as additional flights. Therefore a focus on the coldest homes will limit the growth of discretionary energy consumption. Helping people obtain essential energy services, such as warmth, is good energy economics.
> (Boardman, 2010: 179; and see also pp. 183 and 207)

This is an interesting argument, as it assumes that indirect rebound effects, if they occur, result in more CO_2 emissions than in the direct rebound effects that occur through the consumption of gas or oil in home heating. In fact it is difficult to substantiate this assumption, and no empirical evidence is cited for it. Few empirical studies in this field have been undertaken, and those that exist do not seem to support it. Freire-González (2011) developed an econometric model to estimate direct and indirect rebound effects, using household electricity consumption in Catalonia as an example. He found direct short-term rebound effects of 36 per cent, rising to 49 per cent in the long term. When indirect rebound effects were added to these, the total rebound effects were 56 per cent and 65 per cent respectively. If fuel-poor households show direct rebound effects of around 60 per cent (see above), this would be about equivalent to the total of direct and indirect rebound effects of average households.

In a quite different but related UK study, Druckman *et al.* (2011) estimated indirect rebound effects when households made energy efficient behavioural changes: reducing indoor temperature by 1°C, reducing food expenditure by one-third by eliminating waste, and walking or cycling instead of driving for trips under two miles (3.2 km). Here rebound effects were measured in terms of CO_2 emissions for specific household activities. The average indirect rebound effect was 34 per cent. The authors point out that this compares to 35 per cent for a set of comparable studies in Sweden (Alfredsson, 2000, 2004). However, the study also notes that home heating is one of the most CO_2 intensive household activities compared to other things households spend their money on.

Two points therefore raise questions as to the strength of Boardman's argument. First, there is no compelling evidence to suggest the total rebound effect for average households – direct plus indirect – is significantly different from the direct rebound effect for fuel-poor households. Second, there is no compelling evidence to suggest that the fuel consumption incurred through indirect rebound effects is any more CO_2-intensive than that consumed through home heating. If anything, it is probably less CO_2-intensive.

A further point also needs to be considered. Energy efficiency upgrades in fuel-poor homes are often paid for or substantially subsidised by government or local authorities, whereas upgrades in non-fuel-poor homes are more often financed by the owners. For homeowners this reduces the amount of discretionary spending after the upgrade, as they have either used up a large amount of their savings or increased their monthly mortgage payments. Thermal upgrades very rarely pay their way in terms of fuel savings per amount of money invested (Galvin, 2010, 2014c), so homeowners often have less to spend on extra energy consuming activities after an upgrade than before. This reduces indirect rebound effects for wealthier homeowners who paid for their dwellings' thermal upgrades, but does not have the same effect on the heavily subsidised or free retrofits that are often offered to fuel-poor households.

It does seem, then, that high rebound effects after fuel poverty mitigation are a fact the world has to live with. The intention of upgrading a fuel-poor house is to induce a large rebound effect, and empirical studies to date do not show that average retrofits lead to higher or more CO_2-intensive rebound effects than those in fuel-poor homes.

This does not need to be a problem for policymakers and academics. World leaders have already agreed in principle that rich countries should bear the brunt of climate change mitigation more heavily than poor countries. The basic ethical principle on which this rests ('common but differentiated responsibilities') is that rich countries, which produce far more climate-damaging emissions than poor countries, should pay a greater portion of climate change costs than those who are poorer. Rebound effects are also likely to be high for poor countries which are currently using outdated, inefficient industrial and transport equipment, as they switch to more energy efficient equipment. Rather than try to ignore or downplay the issue of high rebound effects for fuel poverty mitigation, a more productive and credible approach would be to assert the fact of these high rebound effects clearly, while also setting this in the context of globally accepted ethical guidelines such as sustainable development.

Finally, the size of a household's rebound effect is not always a fair measure of the size of its CO_2 emissions. A simple arithmetic oversight seems to run through the discussion on rebound effects in fuel poverty mitigation. A poor household usually has a far smaller 'carbon footprint' than a rich household, as the latter generally travels more (for work and recreation), buys and uses more electronic goods, drives longer distances, eats more imported foods, and takes more holidays in distant places. The quantity of CO_2 a fuel-poor household fails to save after a retrofit probably amounts to about 400kg per year,[3] a very small amount compared to the nine tonnes emitted per year per person in the UK. A rebound effect figure of 60 per cent might look distressingly large, but 4.4 per cent of the average carbon footprint is rather small.

5. Conclusions

This chapter has explored how the rebound effect intersects with the issue of fuel poverty and its mitigation. The two are inextricably linked, because the aim and intention of the mitigation of fuel poverty is to induce large rebound effects in the homes of fuel-poor households. Fuel poverty is characterised by technically inferior dwellings, usually exacerbated by low income among households. The mitigation of fuel poverty has gained traction in policy initiatives in the UK and New Zealand, and is now on the agenda in Continental Europe and an emerging issue in the US (Pivo, 2014).

The mitigation of fuel poverty entails large rebound effects, as evidenced by both empirical studies and graphical modelling. These rebound effects are partly due to technical difficulties with the severe interventions required to bring thermally poor homes up to a decent standard, partly due to the (desired) increase in energy services in these homes, and almost certainly due in part to factors in between, such as difficulties in the interface between occupants and energy efficient technology.

There is a kind of avoidance of discussion of these large rebound effects in literature on the mitigation of fuel poverty, and the logic of this avoidance does not stand up to critical scrutiny. However, the fact that fuel poverty mitigation leads to high rebound effects does not need to cause problems for policymakers and other stakeholders. High rebound effects are inevitable whenever fuel-poor people become beneficiaries of energy efficiency upgrades, in any sphere, as such upgrades are designed and intended to increase their take of energy services so as to enable them to live more healthily and comfortably. The principles of justice and social equity are established in the international context of climate negotiations and human rights. These principles can also be applied within individual countries whereby poorer people have access to a credible minimum level of energy services for health.

Further, although the percentage figures for rebound effects in fuel poverty mitigation look high, they are percentages of very low absolute values. A rebound effect of 60 per cent for a fuel-poor household may represent only a very small shortfall in carbon savings, in absolute terms, compared to the large carbon footprints of the average household in a developed country.

The aim and intention of fuel poverty mitigation is to induce large rebound effects in these homes. This is a worthy and defensible aim, since very little harm but a great deal of good is being achieved through it.

Notes

1   The non-sunny side is not always the north, as the sun is in the north in the southern hemisphere. Fuel poverty happens there as well (see the extensive literature on the New Zealand situation), so it is important to be precise in descriptions of causes of fuel poverty.

2   Readers will recall that the earliest discussion of the rebound effect (even before it was called by that name) focused on its effect in reducing the fuel savings that should have occurred when energy efficiency measures were mandated in response to the oil crises of the 1970s (Khazzoom, 1980). By the turn of the twentieth century, climate change had become entwined in the discussion, as evidenced, for example, in Berkhout et al.'s (2000) seminal paper on the rebound effect in economic theory.

3   Calculated as follows: Suppose a fuel-poor home increases its energy services by 60 per cent after a retrofit and this results in consumption of about $30kWh/m^2a$ more than would have been the case if it had not increased its energy services. If the dwelling has a floor area of $70m^2$, the excess consumption is 2100kW/a. If the heating system uses gas, the annual CO_2 emissions from this excess consumption are 2100kWh x 0.184 kg/kWh = 386.4 kg of CO_2.

7

REBOUND EFFECTS IN NON-RESIDENTIAL BUILDINGS

1. Uncharted territory

Europe's non-residential buildings consume just under 930 terawatt-hours (TWh) of energy per year, compared to around 2440TWh for residential buildings (Economidou, 2011), including heating, lighting and appliances. Residential buildings are of many types but have a single purpose, accommodation, but the diversity in non-residential buildings is vast. They include offices, shops, hospitals, hotels, restaurants, supermarkets, schools, universities, prisons, factories, warehouses and sports centres, and in many cases there are multiple functions in the one building.

The EU Commission's 'low carbon road map' (EC, 2011) aims for a competitive, very low carbon economy by 2050, and the Commission has estimated that greenhouse gas emissions from the building sector will need to be reduced by 90 per cent to fully contribute their share of achieving the specific targets in this plan (Economidou, 2011: 99). There is widespread agreement that improving the energy efficiency of existing buildings is the most economical way to move toward this goal, and this applies as much to non-residential as to residential buildings.

As with residential buildings, there is now a considerable legacy of studies on what is often called the 'technical potential' of such buildings – how to design optimised retrofits so as to achieve desired reductions in their (theoretical) energy consumption (e.g. Juan *et al.*, 2010; Kniefel, 2010; Lam and Hui, 1996). There is no shortage of means for predicting theoretical energy savings through upgrades of non-residential buildings. The problem, however, is that their actual savings are often much lower than predicted (Bordass *et al.*, 2001; Bordass *et al.*, 2004). Energy performance gaps of up to 100 per cent are not uncommon (Hamilton *et al.*, 2011). The rebound effect haunts the non-residential sector as much as the residential sector.

Discussion and research on the rebound effect in residential buildings is well advanced and has produced hundreds of research projects and papers to date, covering a full range of definitions of the rebound effect. This field is much more limited in respect of non-residential buildings. Energy performance gaps (EPGs) are often well documented, and there are now attempts to predict the EPGs that might occur when retrofits take place (Menezes *et al.*, 2011). However, there do not seem to be existing studies on elasticity rebound effects in these buildings.

Nevertheless, there is a tradition of studies of energy saving in non-domestic buildings, often with a special focus on commercial buildings, since 'going green' is becoming an important feature of business activity worldwide (Nelson, 2008; Pivo, 2008). A number of these studies now go beyond the 'technical potential' of energy savings to explore the 'social potential' of such savings. Their key insight is that the human organisation and activity within these buildings has a very big influence on how energy is consumed. Studies such as Janda (2014) and Axon *et al.* (2012) show how this social element can be transformed, in conjunction with the technical potential of a building, to bring significant energy savings.

These studies seek to bridge the gap between 'technical potential' and 'social potential' with a 'socio-technical' framework (MacKenzie and Wajcman, 1985), a notion that has already been discussed in this book (see Chapter 1, Section 3.3).

A difficulty with exploring causes and effects of rebound effects in non-residential buildings is the sheer complexity of factors contributing to energy consumption. A socio-technical framework can offer a simplified analytical model, which provides an ordered perspective on the issue. This is not the only way of approaching the subject, but it provides a navigable pathway through the complexities of the information (for a helpful account of the use of simplified models in scientific research see Harré, 2009).

It is interesting to note that energy saving endeavours in non-residential buildings often seem to naturally develop along lines which are broadly socio-technical. In the UK, the 'PROBE' (Post-occupancy Review of Buildings and their Engineering) project provides feedback for engineers and planners on factors that influence energy consumption and occupant satisfaction in non-residential buildings. Surveys conducted in 1995–2002 and again in 2012 have produced a large body of data and information on how people and building technology interact (PROBE, 2014a). Among the topics investigated are occupant views on indoor temperature, noise level, air quality, air movement, lighting, and personal control over key thermal comfort factors such as heating and lighting (PROBE, 2014b). The process of continual feedback from occupants to engineers is often assumed to be normal in these studies. A picture tends to form of a building as a living entity, where its technical and human members can be influenced to improve their interoperability.

This chapter makes a first attempt to expand this type of approach into discussion of the rebound effect in non-residential buildings, looking mostly at office buildings but also to some extent at other types of non-residential building.[1] As a first attempt in this field, it is exploratory rather than definitive, and includes a degree of creative probing alongside more prosaic description.

Section 2 discusses how the technical features of building fabric and thermal equipment, together with social aspects of occupants and building management can affect energy consumption and therefore rebound effects in office buildings. Section 3 analyses two German datasets of calculated and actual consumption in non-residential buildings. Although these datasets are neither large nor representative, they enable a demonstration of how an estimate of rebound effects could be made as more data becomes available. A Hong Kong dataset is also investigated which does not use calculated consumption as a baseline, but can be used to supplement other results. Section 4 offers comments on possible cross-learnings from these findings for mitigating rebound effects in residential buildings, and Section 5 concludes.

2. Related studies to date

Due to the wide range of functions of non-residential buildings, and variations of usage even within the narrower sector of office buildings, it is not often possible to generalise as to what specific retrofit features would lead to lower rebound effects or lower energy consumption. However, four key issues appear to affect the energy saved and efficiency gains wasted when office buildings are retrofitted: the substance of the building; the involvement of the occupants; the executive actions of building managers; and the quality of monitoring of energy consumption. The first two of these are illustrated in three recent studies from Austria, the Netherlands and the UK, and the last two from a study in Spain. The ways all three interact are shown in a set of building upgrades in London.

2.1 Building substance and energy consumption

Berger *et al.* (2014) investigate how building substance and location characteristics can influence energy consumption in office buildings in Vienna, Austria, given the expected rise in outdoor temperatures due to climate change in the next 35 years. Vienna has hot summers and very cold winters, so that a balance needs to be struck between features that keep an office building cool in summer and warm in winter. Vienna's pre-World War I buildings tend to have thick masonry walls and less window area than modern buildings. This makes them naturally cool in summer but harder to heat in winter. Berger and colleagues find that these buildings perform best when located in inner city heat island zones, consuming less energy overall than when located in cooler, suburban zones. Modern buildings have less thermal mass and thicker insulation, as they are designed to be easy to heat in winter. This gives them an energy consumption advantage over older buildings in winter, but much of this is lost, through an excessive cooling load in summer, due to their large window area with high solar heat gain coefficients and thick insulation which can trap unwanted heat. These summer over-heating effects are expected to increase due to climate change.

Lessons can be learned from this for thermal retrofitting. The most obvious is to install window shades or blinds on the sun-facing sides of a building when insulation is added and windows with high solar gain are installed, preferably with a facility for occupants to control the positions and angles of these shading devices (see below on Meerbeek *et al.*, 2014). Second, there is increasing interest in installing internal wall insulation in older, historic buildings, so as not to compromise the aesthetic or heritage value of the façades. A problem with this is that it robs the interior of the building of the stabilising effect of the thermal mass of the exterior walls. As the interior becomes warm in summer due to human activities, electronic devices and outdoor air incursion, it will no longer be cooled by the thermal mass of the walls, as these are cut off from the indoors by the layer of insulation. Hence a great deal of care needs to be taken when considering whether and how to insulate one of these thick-walled masonry buildings. Retrofits could also be optimised to suit the local climate zone, depending on whether this is an inner city heat island, an equator-facing rural slope, or something in between.

2.2 Effects of occupant behaviour

Occupant behaviour issues can be somewhat different in office buildings than in homes. This can in part be due to the lack of energy savings incentive, as the workers do not pay the fuel bills. In economics literature the rebound effect is considered entirely as a 'price effect' (Berkhout et al., 2000; Sorrell and Dimitropoulos, 2008), whereby occupants use a higher level of energy services because this is now cheaper, due to increased fuel efficiency. But if there is no personal savings motivation for reducing energy consumption, there can be no rebound effect by this definition. In recent field work by this author (as yet unpublished) a significant number of office workers are saying they do take this attitude: while being careful to save energy in order to save money in the home, they do not apply this reasoning at work. As well as pointing to the need for a broader definition of the rebound effect for office buildings, this also raises important questions as to how to motivate workers to participate in energy savings regimes in the office.

There is also the socio-technical issue of how space is apportioned in office buildings compared to homes. Many office buildings are now open plan, so conditions can vary widely within these large spaces. Those next to the windows will have different experiences in terms of thermal comfort from those more distant from them. So it can be a problem to optimise the comfort of different groups of people who are experiencing different indoor conditions due to their spatial position within the building.

A further socio-technical issue is the interactions of occupants with energy saving technology. Meerbeek et al. (2014) report an investigation in Eindhoven, Netherlands, of how office workers controlled exterior blinds which had been installed for managing the amount and timing of solar gain in their offices. The blind positions on the south side of 40 offices were photographed at six minute intervals over 21 weeks from late July to mid-December in 2011, a period including summer, autumn and winter conditions. Seventeen of the office workers were selected to keep diaries of how they perceived the indoor temperature, light and air quality, and were interviewed on these issues and their blinds usage. Blinds usage was recorded alongside precise, hourly weather data for the observation period. The blinds could all be operated manually by the office workers. Some had automatic controls but these had override facilities, which were frequently used. Only one-quarter of the blind adjustments in the research period happened automatically, the rest by occupant intervention.

Results showed that office workers made manual use of the blinds in response to sunlight, cloud cover, and outdoor temperature. Individual office workers had different preferences for sunlight or shade, and being able to adjust the blinds manually gave them a sense of being in control, which led to satisfaction with their indoor environment even when it was not ideal for their preferences. Their blind adjustment routines did not always square with the computer simulations of maximum comfort and energy saving which drove the automatic blind adjustment systems, but being able to override the automatic system seems to have helped win the occupants' cooperation as partners in energy saving.

An example of a simple way of engaging the support of workers in office buildings is offered by Tetlow et al. (2014). This UK study engages with the perennial problem of office workers not switching lights off after rooms are vacated, even among those who are highly motivated to save energy. The authors framed this as a 'post-completion error', similar to forgetting to attach a document to an email before sending it. Such errors are said to occur because the outcome of a task is achieved before all the intended actions are accomplished (for a fuller exposition of the psychology of such errors see Byrne and Bovair, 1997 and Chung and Byrne, 2008). Tetlow and colleagues hung very visible, carefully designed cue signs alongside light switches in meeting rooms in office buildings, and found a 54 per cent increase in the incidence of switching off lights at the conclusion of meetings (from 56 to 86 out of every 100 room uses). Further, by leaving the signs in place they found the new switch-off rate did not diminish after six months. This provides a small but clear example of a mechanism through which office workers can become effectively engaged in energy saving at work.

A different example of engaging worker support is reported by Ackerly and Brager (2013). This study investigated how workers in office and mixed use buildings responded to signalling systems which advised them when to open or close windows. Using 16 case study buildings, the study examined both the design of the signalling/opening system, and the ways workers interacted with it. Suggestions were then made as to how trade-offs and synergies between energy saving and occupant comfort can be improved, through technical design and worker engagement strategies that take the social and the technical into account in an integrated way.

2.3 Building management and energy monitoring

The role of building management is highlighted in a study by Vázquez (2013) of three office buildings in the Basque area of Spain. These buildings were owned by Tecnalia R & I, a private research institution, and situated in Donosti (two buildings) and Bilbao. For a given set of occupant behaviours, management systems and thermal technology, Vázquez found the largest determinant of energy consumption to be the outdoor temperature, a finding echoed in other studies of non-residential buildings of various types (e.g. Katunsky *et al.*, 2013).

Within that constraint, however, the biggest influences over energy wastage and potential savings were building management and energy monitoring systems. The most common cause of energy wastage was extended running periods of heating, ventilation and air conditioning (HVAC) systems, and these were decided in each case by building operators. Building operators were not aware of energy consumption figures, as they did not manage the energy bills. They assumed that if occupants ceased complaining, the HVAC systems were operating optimally. Therefore they set the systems to run for longer periods than necessary so as to avoid complaints. Further, they only checked the operating efficiency of components in the HVAC system when there were complaints relevant to these, so that non-optimised components could remain faulty for long periods without being noticed. This could be framed as a socio-technical failure, in which the human organisation and systems in the building did not harmonise optimally with the technical components.

Further, there were no detailed metering facilities to monitor gas consumption in the two buildings in Donosti, though in the building in Bilbao these facilities were available. However, the systems for storing, retrieving and acting on the detailed data these produced were not well developed. Also, in some cases the electrical energy that drove the fans to distribute warm or chilled air throughout the building was not accounted for as heating energy consumption, so that the energy consumed for heating and cooling was underestimated by building owners.

Although the building owners were concerned about energy efficiency, there seemed to be an assumption that this was entirely determined by the HVAC and other thermal technology. Energy monitoring and management were poorly established and ineffective. The result was that energy was consumed without proper monitoring and control, and opportunities for energy saving were missed. Again this can be conceived as a socio-technical breakdown.

An important implication of this study is that building management needs to be actively engaged in energy saving if this is to happen, providing building operators with sufficient tools to manage and influence energy consumption, for example by setting appropriate defaults, and having mandates strong enough to do more than merely respond to complaints. Goins and Moezzi (2013) propose an enhanced complaints handling system for office buildings that goes beyond merely responding to each reported problem as it arises. Instead it sets complaints within the social and technical context of the building and uses logging and analysis, so that management's response becomes part of an ongoing commissioning of the building.

Some of the ways factors such as these can be turned to energy saving when energy efficiency upgrades are undertaken in office buildings are illustrated in the following five projects.

2.4 Optimised retrofits in UK office buildings
The commercial property management firm Jones Lang LaSalle (http://property.joneslanglasalle.co.uk) recently implemented comprehensive energy saving strategies in five of its clients' London office buildings (Jones Lang LaSalle, 2014 and links). These are of particular interest to rebound effect studies, as they combine technical energy efficiency measures with finely tuned building management and the development of cooperative occupant behaviour to save energy. The main energy saving measures, savings in energy and CO_2, and costs of these five properties are listed in Table 7.1.[2] It should be noted that this information is taken at face value from the firm's own publications and have not been peer-reviewed.

The energy and CO_2 savings shown in Table 7.1 are considerable, and suggest that rebound effects, if any, were quite small after these energy efficiency improvements. Costs of refurbishment ranged from a low of £7/m^2 of floor area to a high of £114/m^2, with an average of £46/m^2 for the four cases where this is given. This included water and waste disposal measures, so the cost for thermal improvements alone would have been even lower. This compares with costs of £150–£500/m^2 for typical, comprehensive thermal retrofits of houses (Galvin, 2010; Jakob, 2006).

Several further points are noteworthy. First, the technical upgrade measures were almost all quite small-scale but very carefully targeted. There were no insulation measures and few window replacements. These buildings are generally thick walled, with plenty of thermal mass, and in one case (Hollywood House) extra thermal mass was introduced inside the building to stabilise

Table 7.1 Integrated energy saving measures developed by Jones Lang LaSalle in five London office buildings, as reported in this firm's publications.

Building	Energy saving measures	Savings and cost
Hollywood House, SW15 3RN. 5-storey, 4,035m². Multi-firm occupation	• Photovoltaic and solar thermal panels; • Connection to District Heating; • Feedback metering of electricity, heating and water with warnings when benchmarks exceeded; • High efficiency lighting, with automatic dimming; • Ventilation system integrated with window opening and sensing of human presence; • Automatic switch off of equipment out-of-hours, with override; • Extra thermal mass in raised floors and lowered ceilings.	Energy consumption reduction 56%; CO_2 savings 113 tonnes/yr ($28kgCO_2/m^2$ of floor area/yr) Annual savings £28,000; Capital investment £460,000 ($114/m^2$).
10 Exchange Square, EC2A 2BR. 16,100m². Multi-firm occupation including a Green Building management Group	• Reduced daily running times for heating and cooling; • Disabling of out-of-hours overrides (manager now decides on out-of-hours energy use); • Motion sensors for lighting in occupied areas; • Fuel conditioning fitted to gas supplies.	Energy consumption reduction 35%; Annual savings £111,050; CO_2 savings 765t/yr ($95kg/m^2a$, or Capital investment £74,000 [$46/m^2$]).
40 Grosvenor Place, SW1 7AW. 6-storey, 23,000m². Multi-firm occupation	• 143 sub-meters fitted to monitor heavy appliance use; • Hourly access to real-time consumption data for all occupiers and building managers; • Lighting audits to identify savings potentials; • Daylight sensors and motion sensors installed; • Cooling system now utilizes cool outdoor air when appropriate; • More efficient air quality control making fuller use of indoor fresh air; • Change to high efficiency lighting; • Removal of automatic temperature monitors; team now matches heating and cooling controls to office hours and indoor–outdoor temperatures; • Colour-coding in occupiers' energy bills to help identify potential savings.	Energy consumption reduction 38%, or $329kWh/m^2a$; CO_2 savings 3,530t/yr ($153kg/m^2a$); Annual savings £135,000; Capital investment £151,000 ($7/m^2$).
Prospect House. 3-storey, Single firm-occupation.	• Introduced a live environmental log for recording and disseminating progress and initiatives on green issues; • Energy audits to identify potential savings; • Reduced boiler and hot water capacity to match reasonable demand; • Adjusting plant running times to match occupancy; • Running 'energy weeks' every six months to engage with occupiers and raise awareness.	CO_2 emission reductions 15%, i.e. 279t/yr; Annual energy savings £50,000.
50 Pall Mall, W10 6BL. 7-storey, 3,362m². Multi-firm occupation.	• Reconfiguring reception area layout and installing double doors for air tightness and to reduce heat loss; • Installing motion sensors in the stairwells and on fourth and seventh floors to control lighting in small zones; • Replacing boilers with models up to 30% more efficient; • Installing energy efficient lighting; • Permanently switching off atrium lights.	Energy consumption reductions: gas 54%, electricity 22%; CO_2 reductions 284t/yr ($84kg/m^2a$); Annual savings £41,500; Capital investment £60,000 ($18/m^2$).

Source: Jones Lang LaSalle, 2014 and links

temperatures. Second, many of the measures have to do with timing. The timing of ventilation, lighting, cooling and heating periods was matched to actual usage times within the buildings. In some cases this was achieved by automatic cut-outs with overrides, but in one case existing overrides were removed so that building operators could make executive decisions on out-of-hours energy consumption. Third, natural heat, coolness and air quality were utilised alongside artificially-induced temperature change and air conditioning. Fourth, many of the changes were made after thorough audits were carried out to identify savings potential, rather than simply making changes that are generally regarded as energy saving.

Finally, the users of the buildings were engaged with the upgrades in a dynamic way. This is seen to some extent in Table 7.1, for example in the keeping of an environmental log book (Prospect House), access for all occupants to real-time consumption data (40 Grosvenor Place) and regular 'energy weeks' for occupants (Prospect House). However, the full extent of user involvement is more clearly seen in Jones Lang LaSalle's (2014) overview of how the upgrade projects were conceived. They began with a process of consultation and discussion aimed to engage the building owner, the firms whose business premises are in the buildings, and the people who work in them. Decisions as to how the retrofit would proceed emerged from these discussions, as did the management strategies for aligning the human and technical features of the buildings, so as to optimise savings.

In this respect the notion of a 'socio-technical system' is useful for understanding the approach to buildings and retrofitting in these examples. As discussed in Chapter 2, socio-technical systems theory suggests it is a mistake to see human society and technology as two separate spheres. Rather, technology shapes society, just as society shapes technology, and it is impossible to understand one without acknowledging the role of the other. Hence a building is not just a physical, technical entity which people come into to occupy and operate. Rather, a building and its occupants are more like an organism, which functions best if all the constituent elements fit together well and mutually support each other's needs. Seen this way, it would be a mistake to design a retrofit of an office building simply according to theories of optimum energy saving, so as to achieve its 'technical potential'. Far more can be gained, for a far lower cost, if the approach is to tweak the building and engage the occupants so that the socio-technical system of people and material, technical fabric move forward together toward greater energy savings. The technical and social potential of a building need to be optimised in harmony with each other.

It must again be emphasised that the Jones Lang LaSalle reports have not been subjected to the rigour of peer review and are taken here on trust. Nevertheless, they are publicly available documents produced by a firm which needs to protect its reputation.

2.5 Hypotheses on reducing the rebound effect in office buildings
Insights from the studies outlined above can lead to a set of broad hypotheses as to how rebound effects could be reduced in office buildings:

- Rebound effects may be lower when technical upgrade measures are optimised to suit local thermal conditions, the thermal characteristics of the existing building substance, and types of energy saving that can be enhanced by appropriate building management and active occupant engagement – rather than comprehensively retrofitting every thermal feature of every building to achieve the highest possible theoretical energy savings.

- Planning a retrofit should involve wide-ranging discussions with all levels of personnel who will be using and managing the building, in an open exchange of ideas. It is especially important to get the occupants and building operators on side and learn from their localised experiences within the building.
- Bigger is not always better with retrofits. Insulation is not always appropriate and needs to be used carefully. Boiler and cooling systems, and hot water storage, might need to be down-sized.
- Retrofits need to include consumption monitoring facilities which are detailed enough to provide useful feedback which can be acted upon to tweak systems and behaviours for more efficient thermal performance. Proper management systems need to be implemented to make optimum use of this data.
- The timing of heating, cooling and lighting periods is important in energy saving. Systems need to be designed so that for each building and its uses the most suitable combination of automatic controls, overrides, and building operator control can be developed when the retrofitted building starts operating. Natural ventilation for a portion of the year may also be an option.
- The completion of the retrofit is not the end of the process. A new phase then begins, involving active energy management and the cooperation and input of building occupants.
- Occupants generally need to feel they have a degree of control over their thermal environment, and may support savings measures more enthusiastically if they are allowed to experiment and make mistakes, rather than being confined to prescribed levels of temperature, solar incursion and humidity. Insights from literature on 'adaptive comfort' can be helpful in this regard (Nicol and Humphreys, 2009).
- In general it is useful to let the retrofit strategy be guided by the notion of the human and technical-material elements of the building in an integrated socio-technical system.

3. Consumption datasets and rebound effects in non-residential buildings

Currently there are no suitable datasets of calculated and actual consumption in non-residential buildings prior to and after thermal retrofits. However, there are two datasets, both from Germany, of calculated and actual heating consumption of non-residential buildings in their current state. One of these, from the environmental and housing research institute Institut Wohnen und Umwelt (IWU) in Darmstadt, gives data for 93 buildings of various types (IWU, 2014). Another, from the German Energy Agency (*Deutsche Energie-Agentur* – DENA) gives data for 32 office buildings (Oschatz *et al.*, 2014). Using the cross-sectional method described in Chapter 4 it is possible to estimate likely average rebound effects for buildings in datasets of this type, assuming that if they were retrofitted their calculated and actual consumption would, on average, move leftward along the same curve. In comparison to the thousands of such data points for residential buildings in Germany and the 913 for France, these datasets are small, so the margins of error in the estimates below are very large. They give only the broadest idea of what kind of average rebound effects could be expected if these buildings were retrofitted, but they also provide some important qualitative insights. A further dataset, from Hong Kong, is also small and has different parameters but can add to the findings here.

3.1 A small dataset of office buildings

The smaller dataset, from Oschatz *et al.* (2014) is the most homogeneous as it consists entirely of office buildings. This is displayed in Figure 7.1.

The straight broken line in Figure 7.1 represents the points where actual and calculated consumption would be equal. As with residential buildings the line of best fit for the data points is a power curve, in this case given by the equation:

$$E = 3.5281C^{0.7078} \tag{7.1}$$

Recalling from Chapter 4 and the Appendix, Section A.4, that the elasticity rebound effect is 1.0 minus the exponent of this equation, the elasticity rebound effect here is:

$$R = 1 - 0.7078 = 0.2922 \ or \ 29.22\% \tag{7.2}$$

An elasticity rebound effect of 29.22 per cent means that for every energy efficiency increase of 1 per cent, 29.22 per cent of this efficiency gain is lost to increased energy services, technical failures, etc., while the remaining 70.78 per cent is utilised to reduce energy consumption.

If this were a large dataset, this would mean that the average rebound effect when a number of these buildings are retrofitted would be approximately 29.22 per cent. However, there are not enough buildings in this dataset to make such a bold assertion. Instead, all that can be said is that there are striking similarities between the general shape of this data plot and those for residential buildings, and that the curve for these office buildings has a much higher exponent, indicating that their rebound effects appear to be much lower.

It is also informative to observe how the EPG varies for different values of calculated consumption. A plot of this is given in Figure 7.2.

Figure 7.2 indicates that, as with residential buildings, the office buildings with the lowest calculated consumption (the highest energy efficiency) have the highest EPGs. The better the building, the more over-consumption tends to occur. Further, these EPGs are much higher than those for residential buildings, i.e. they have smaller negative numbers and larger positive numbers. The buildings' average EPG is −0.153. The average prebound effect, which is the numerical negative of the EPG, is 0.153 or 15.3 per cent, compared to around 35 per cent for German dwellings and nearer to 40 per cent for those in France. This indicates that these office buildings perform more closely to their calculated ratings. This implies that, although their rebound effects are low, this might not represent much of an advantage because their high EPGs indicate they are already big consumers in relation to their calculated energy efficiency.

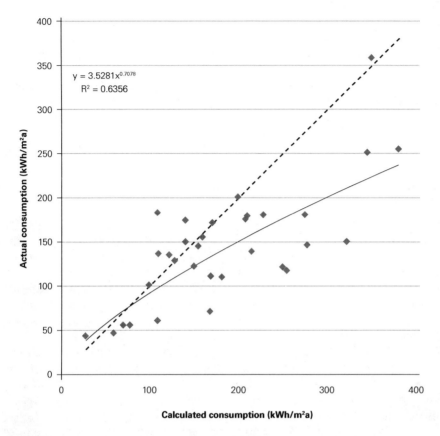

Figure 7.1
Actual and calculated heating consumption for 32 office buildings, with best-fit power curve
Data source: Oschatz et al. (2014)

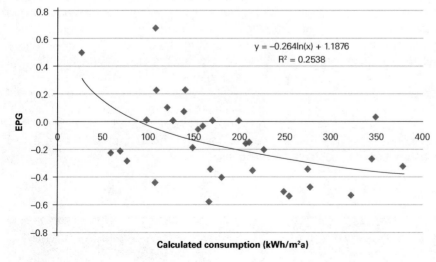

Figure 7.2
Energy performance gap, 32 office buildings
Data source: Oschatz et al. (2014)

3.2. A larger but mixed dataset

The dataset from IWU (2014) is almost three times as large as that from DENA, but consists of a number of different types of buildings. These are given in Table 7.2.

A plot of calculated against actual heating consumption for these buildings is given in Figure 7.3.

Here the line of best fit is given by the equation:

$$E = 1.8803C^{0.8292} \tag{7.3}$$

The elasticity rebound effect is therefore:

$$R = 1 - 0.8292 = 0.1708 \ or \ 17.08\% \tag{7.4}$$

This would mean that in energy efficiency upgrades, only 17.08 per cent of every 1 per cent increase in energy efficiency would be lost to increases in energy services, technical failings, etc., while 82.92 per cent would go toward reducing energy consumption. As with the previous dataset of 32 office buildings, the rebound effect is much lower than is typical in residential buildings.

The EPGs of these buildings are plotted in Figure 7.4. Again the EPGs are high compared to those of residential buildings, with an average of −0.165, or an average prebound effect of 16.5 per cent. As with the previous dataset they are also higher for the more efficient buildings, i.e. those with lower calculated energy consumption.

There is an interesting and important extension to this set of data. The IWU research team used two different methodologies to find the calculated consumption figures for these buildings. The official German method, published by the German Institute for Standards (*Deutsches Institut für Normung* – DIN) as DIN-18599, was used for the calculated consumption figures given in Figure 7.3. This method is used in building regulations to guide architects and engineers in ensuring their building designs conform to national requirements. It indicates how to calculate the heating energy that would be required to give full thermal comfort in a building of a particular size, shape and thermal componentry. 'Full thermal comfort' is defined as having a constant indoor temperature of at least 19°C in all rooms all year round, with an adequate level of ventilation. Any building with its data point lying on the broken line in Figures 7.1 and 7.2 would therefore be operating at a level to give full thermal comfort according to the DIN-18599 specifications, as this is the line where actual consumption is equal to calculated consumption.

Table 7.2 Types of buildings included in dataset of actual and calculated consumption from IWU (2014).

Conference and event centres	Sports centres	Office buildings	Other commercial uses	University buildings	Hotel, hostels	Schools and kindergartens
16	1	23	11	19	8	15

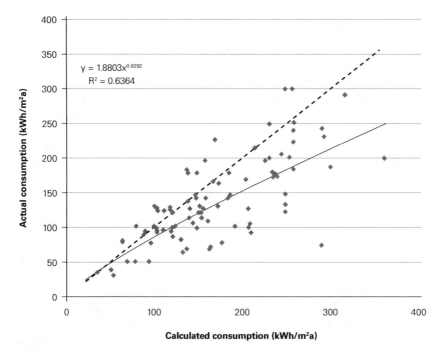

Figure 7.3
Actual and calculated heating consumption for 93 office buildings, with best-fit power curve
Data source: IWU (2014)

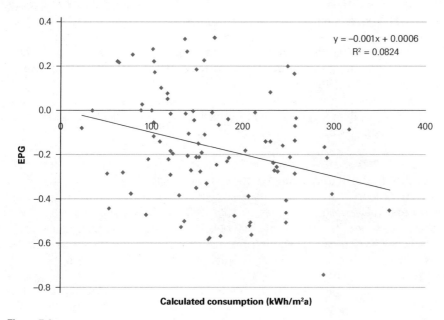

Figure 7.4
Energy performance gap against calculated consumption for 93 office buildings
Data source: IWU (2014)

The second method the research team used was derived in an attempt to produce calculated consumption figures that are closer to actual consumption, i.e. that represent real building situations more closely. These figures shift most of the data points in Figure 7.3 horizontally toward the left. This increases the EPG for all the buildings (gives it smaller negative values), *but does not affect the elasticity rebound effect significantly*. For example, if the new calculated consumption figures are each 0.85 of the original DIN figures (which is approximately how the research team scaled them), the average EPG rises from –0.165 to –0.017. The elasticity rebound effect, however, stays the same, at 17.08 per cent. This is because the elasticity rebound effect is based on a comparison between pre- and post-upgrade ratios of actual and calculated consumption, rather than merely on a single data point of actual and calculated consumption (for a mathematical proof of this see the Appendix, Section A.6). This reinforces the point that *the elasticity rebound effect is a very robust measure*. The new scale of calculated consumption developed by the IWU researchers is effectively a re-scaling of the energy efficiency of these buildings, as energy efficiency is the reciprocal of calculated energy consumption (see Chapter 1). Regardless of what scale is used for energy efficiency ratings, the rebound effect will always come out the same. The rebound effect measures how much worse a building-plus-occupants performs after an energy efficiency upgrade, *in proportion to* how they performed before it. Re-scaling building efficiency measures does not affect this calculation.

Perhaps the most striking finding from these two datasets of non-residential buildings is that their elasticity rebound effect is significantly lower than that of residential buildings while their EPGs are significantly higher. Having lower rebound effects means that performance does not change much when energy efficiency is increased – it does not get much worse than it was before. But having high EPGs means that performance was not especially energy saving in the first place. It can therefore be useful to keep both measures in mind when assessing energy performance and energy saving, as they can complement each other to give a fuller picture.

Nevertheless, it must be emphasised that these are small datasets which might not well represent the performance of non-residential buildings in general. The can only be used as a preliminary guide.

3.3 Consumption and building age in a Hong Kong dataset

Chung *et al.* (2006) developed a methodology for specifying which factors lead to what proportions of energy consumption, in supermarkets in Hong Kong. The weather-adjusted actual consumption of 30 such buildings is given in their study, together with the age of each building. Consumption here includes all energy sources: heating, lighting, refrigeration, etc. A plot of this data is reproduced in Figure 7.5. The horizontal axis is age of building rather than calculated consumption.

Energy consumption in supermarkets is very different from that in office buildings, particularly because of the large amount of refrigeration. However, with all types of building, age generally has a relationship to energy efficiency. Older buildings are likely to be less energy efficient in their thermal qualities and electrical plant, than newer buildings. This would not be the case, however, where equipment has been recently upgraded, which is much easier to do for

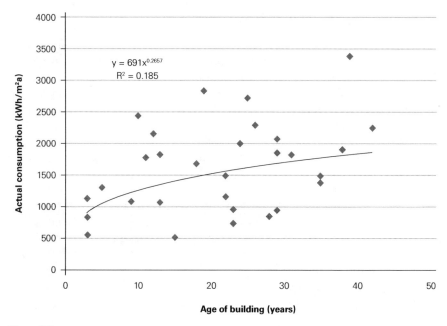

Figure 7.5.
Age and weather-adjusted energy consumption of 30 Hong Kong supermarkets
Data source: Chung et al. *(2006)*

supermarket refrigeration than for building substance. This might help explain why the vertical spread of data is so great in Figure 7.5. One 14-year-old supermarket is performing as well as the best one-year-old supermarket; two supermarkets which are almost 30 years old are performing better than one of the one-year-old supermarkets; and one of the worst performing is just 10 years old.

A further point of interest is the shape of the best-fit curve in Figure 7.5. A range of mathematical functions were tried (logarithmic, linear, polynomial, power and exponential), but again a power curve has the best fit, indicated by the highest R^2 (correlation) value. If the age of a building is roughly related to its thermal quality, the shape of this curve would support the finding of Sections 3.1 and 3.2 above, that cross-sectional data such as this could be used to estimate average rebound effects, if a dataset is sufficiently large.

The exponent of this curve is 0.2657, a low value compared to those for other non-residential buildings, suggesting modernisation might not lead to such high energy savings in supermarkets as it would in office buildings. However, there are too many unknowns here, and the dataset is too small, to make confident generalisations.

4. Lessons for rebound effects in residential buildings

Residential buildings are different from non-residential buildings in ways that relate to energy consumption and rebound effects. Residential buildings can be much smaller than non-residential, therefore they have a higher ratio of surface area to volume and lose heat more readily, thus benefiting more from thick insulation. Where residential buildings are as big as or bigger than non-residential buildings, they are divided into autonomous apartments and usually do not have an overall, building-wide energy management system. A residential building's heating system cannot be turned off at 6pm on weekdays and for the weekend, and the coming and going of occupants is irregular and unpredictable compared to the uniformity of office hours. Within each household there are often poor systems of energy management, and the hierarchical structure of the workplace is less appropriate to families and even less so to shared households of autonomous individuals. Further, most non-residential buildings have specific purposes, leading to indoor routines which are predictable and do not undergo sudden major changes from time to time, whereas such changes happen in dwellings when occupancy changes or when occupants' daily routines change, such as when an occupant changes jobs or becomes unemployed. The main energy demand for office buildings is cooling (not heating) due to the density of people and equipment. Therefore it is less risky to design a non-residential retrofit for a specific set of indoor activities, but there may be considerable risk in designing a residential retrofit on the assumption that the unique lifestyle currently occurring in the dwelling will continue long term.

Domestic retrofitting can nevertheless observe some useful lessons from successes and failures in the non-residential sector.

- Retrofits can sometimes be planned with actual living patterns in mind rather than one-size-fits-all. For example, an elderly infirmed occupant might benefit from underfloor heating or extra thermal mass built into the floor to stabilise indoor temperatures, but this may be wasteful for a working household where members come and go frequently.
- Like office workers, householders need to feel in control of their indoor environment and be allowed to make mistakes. Controls should be extremely user-friendly and easily accessible, including thermostats, timers, ventilation and sun shading, on a central and room-by-room basis. The system needs to give feedback to indicate when it is not optimised.
- Thought should be given to systems of energy consumption management when retrofits are planned. If a household is weak in management skills, there may need to be more automation and less control interfacing. Households with good management regimes might cope well with finer controls and more override options.
- In some cases the greatest energy savings and lowest rebound effects will be achieved when an upgrade is planned to meet the specific needs of a particular household and their particular dwelling. An expensive retrofit with external wall insulation, floor and ceiling insulation, new windows and a new heating system may be appropriate for one building–household combination, but inappropriate for another. In other cases good results might be achieved through small, well-targeted measures: for example insulation of one very cold wall; replacement of a draughty front door; installing thermal curtains; thicker loft insulation; a more user-friendly thermostat and timer; sealing draughty cracks in walls and around windows; removing an evergreen tree on the sun-facing side of the house.
- Thermal upgrades need to be accompanied by discussion on the motivation for saving energy. For many households, the cost of energy is not a great motivator to adopt a stringent energy saving regime, since energy bills are

still a small part of most household budgets (except for fuel-poor households). A large part of the motivation for the London upgrades (Section 2.4) was environmental concern, as this is becoming a major plank of many businesses. Although these upgrade measures paid back in just a few years through energy savings, the energy consulting firm that led the upgrades also promotes its work as green and climate friendly, and this can act as a motivator for gaining the support of occupants. For households to take the trouble to engage daily with heating controls and ventilation routines, they need to have strong motivation. Planning a retrofit could be a useful opportunity to engage a household in discussion of this issue.

- As with non-residential retrofits, framing a residential retrofit as a socio-technical system change can be a useful guiding principle for retrofit strategy and planning. This resonates with recent findings regarding residential buildings (see, e.g. Gram-Hanssen, 2014; Vlasova and Gram-Hanssen, 2014 and literature review in Galvin and Sunikka-Blank, 2014a).

5. Conclusions

This chapter has offered a first attempt to explore rebound effect issues in relation to non-residential buildings, with most emphasis on office buildings, set within a 'socio-technical' framework. The chapter has brought together studies and data which are relevant to this theme, and drawn some preliminary conclusions as to what is likely to influence rebound effects in energy efficiency upgrades of these buildings.

Four main factors appear to play a major role: the physical substance of the pre-upgrade building; the roles and engagement of occupiers in energy consumption; the role of day to day building management; and the advantages of detailed consumption monitoring with good retrieval and consequent actions taken. It is also important to plan upgrades in consultation with all those who will be affected, especially occupants and building operators, and to optimise likely savings rather than simply apply all possible thermal improvements. The notion of a socio-technical system is a useful framework for approaching the planning of non-residential retrofits, as well as the analysis of rebound effect determinants.

Five London case studies were examined as a set of examples of what can be achieved. Although the data and information on these has not been peer-reviewed, the findings as such seem credible and are available to public scrutiny. Significant energy savings were achieved at low cost, through strategic targeting of upgrade measures and system changes to suit particular buildings and their occupants and uses, set in what is effectively a socio-technical systems approach.

Two datasets of calculated and actual consumption were examined using the cross-sectional method of estimating rebound effects (outlined in Chapter 4). One of these consists of 32 office buildings and the other of 93 non-residential buildings of a range of types. Although these datasets are small, the same phenomenon was observed for these as for the larger datasets of residential buildings examined in Chapter 4: the line of best fit is a power curve, from which likely average elasticity rebound effects can be calculated. These rebound effects are significantly smaller than those for residential buildings. However, their EPGs are significantly larger. This would indicate (if the datasets were large enough to justify statistical inferences) that although performance does not get much worse when these buildings are upgraded, it was energy intensive to start with, compared to that in residential buildings.

A further, incidental observation was that the elasticity rebound effect does not change when the calculated consumption is derived through linear scaling methods which bring it closer to the actual consumption.

Finally, the question of common lessons for non-residential and residential buildings was briefly explored, for minimising rebound effects. Although there are significant differences between the two sectors in regard to the buildings' physical characteristics, usage patterns and management regimes, some lessons can be learned for residential retrofits. These concern retrofit planning, energy savings motivation, matching retrofits to building–household characteristics, and matching thermal controls to management skills. Again it was suggested that the notion of a socio-technical system can help guide retrofit planning. Some recommendations for research, policy and planning are:

- There would be much to be gained from framing residential retrofit policy and practice in a socio-technical context. Both the technical and social potential of retrofits need to be explored together, keeping their inextricable nature in mind. In many ways the commercial building sector is ahead in this, though this sector often has the advantage of a clear, functioning organisation within their buildings.
- A key element in this is full consultation with stakeholders at the early planning stage of a retrofit. It is not simply a matter of offering an array of technical upgrade measures on a take it or leave it basis. With incentive projects such as the Green Deal, this seems to be the norm. The energy advice offered at this early stage needs to be fully cognisant of both technical and social/behavioural factors in energy saving, and also the ways these interact and can be developed to optimise savings and occupier satisfaction. The technical team cannot determine these points without the occupants' inputs.
- For energy and emissions policy planning, elasticity rebound effects need to be considered in concert with EPGs. A low rebound effect does not necessarily signal that all potential savings are being realised. If the EPG is still high in comparison with averages for the same type of building, there is likely to be a lot more scope for savings through behaviour change.
- While studies of behaviour and organisation in energy consumption in commercial buildings are well advanced, and EPGs are often documented, there are still no known robust studies on elasticity rebound effects in this sector. Such studies need to be fostered and supported.

Notes

1 There is also a growing number of studies of energy consumption and the EPG in schools (Hong *et al.*, 2014), but as this is developing into a separate and specialised topic, it would need to be covered in depth in a separate study.

2 Refurbishment and reorganisation also covered water usage and waste disposal, but these details are not included in this discussion nor in Table 7.1.

8

CONCLUSIONS, INSIGHTS AND RECOMMENDATIONS

1. Introduction

This book has confirmed that the rebound effect in domestic heating is a major and pressing problem for policymakers and others concerned with energy planning and consumption. Average rebound effects in domestic heating appear to be in the range 20–50 per cent for western European countries. The cross-sectional method, which gives the most reliable results, suggests average rebound effects of 36 per cent for Germany and 50 per cent for France. This means that when homes are upgraded in Germany, only about 64 per cent of the increase in energy efficiency actually leads to reductions in energy consumption. The remaining 36 per cent is lost to the rebound effect. In France this could be as high as 50 per cent. There is no reason to believe rebound effects in other western European countries are far outside this range.

The prebound effect shows that people in older, less energy efficient homes get used to living with less heat than is theoretically required to provide comfortable indoor temperatures in all rooms, all year round. Some of this is due to fuel poverty and needs to be remedied, but mostly it seems that people have a wide range of adaptation to indoor climates. However, it is clear that this advantage is lost, on average, when homes are retrofitted to high thermal standards. In these 'low energy' homes, energy performance gaps (EPGs) are consistently high, and rise steeply as the thermal standard increases. In passive houses and the conventional low energy homes for which data is available, EPGs of around 45 per cent are typical. These homes are consuming almost half as much energy again as the amount they were designed to consume.

Not only does this waste energy, it also compromises energy and CO_2 emissions reduction targets. If policymakers wish to see energy consumption reductions of 80 per cent by 2050, the average energy efficiency increase in the building stock will need to be significantly higher than 80 per cent, due to the rebound effect.

There appear to be promising solutions to some key aspects of this problem, but by no means all. This book has suggested adopting a 'socio-technical' approach to thermal retrofitting and energy saving. Energy can be saved by upgrading the physical and technical features of a dwelling *together with* the active engagement of its occupants. If this approach is taken, from the planning stage through design, implementation, feedback and post-retrofit support, the potential for savings seems to rise significantly. This is highly unlikely to bring the 80 per cent reductions sought by policymakers, but it could go further toward realising the savings potential of each specific dwelling.

This chapter brings together some of the more important insights that have emerged throughout the book, suggesting how they can lead to recommendations for various stakeholders. These stakeholders include those who are directly and indirectly affected by rebound effects in domestic heating, such as homeowners, tenants, housing providers, the building industry, policymakers, energy suppliers, planners and other civil authorities. There are also stakeholders in the academic community, as their research and conceptualising of the rebound effect can help others deal better with rebound issues. This chapter is structured in terms of insights, most of which lead to recommendations which may apply to one or a number of stakeholders in different ways. Some of these insights are of a more abstract or conceptual nature but, nevertheless, have practical as well as theoretical applications. Others bear directly on the practicalities of minimising rebound effects when energy efficiency increases occur.

2. A dwelling-plus-occupants as a socio-technical system

A helpful way to conceive of a home in relation to energy consumption and efficiency is as a socio-technical system (Chapters 1, 2 and 7). The determinants of energy consumption in a home cover a wide range. Dividing them neatly into human influences and material-technical influences is not always helpful, and can give a skewed picture of what is causing what to happen. Empirical research on homes after energy efficiency upgrades reveals a spectrum of mixed reasons as to why these homes usually do not achieve the energy consumption reductions that were calculated and expected. These include the 'price effect' of taking more thermal comfort because it is now cheaper to do so; the thermal effect of the dwelling simply being warmer no matter what the occupants do; the difficulties of operating over-complicated thermostats and timers; mismatches in the hourly and daily rhythms of heating elements and occupant lifestyles; mismatches between different pieces of retrofit technology, some of which occur because of further mismatches with occupant lifestyles; technology failures such as poorly applied insulation; and bad planning, as when positive thermal effects of thick masonry walls are lost through over-zealous insulation.

Conceiving a home as a socio-technical system provides a framework which can take all these elements into account as they are, and deal with them in a more comprehensive way. This leads to the following recommendations:

2.1 Economists need to re-think whether the 'rebound effect' is entirely a 'price effect'

Studies of what came to be called the rebound effect began with economists, who brought excellent conceptual and mathematical tools to theorise how people respond to price changes. A basic contribution of early theorists such as Khazzoom (1980), Brookes (1990) and Berkhout et al. (2000) was to show how energy efficiency upgrades made the effective ('endogenous') price of energy lower, leading to increases in the consumption of energy services – the phenomenon economists came to call the rebound effect. This led to the development of a comprehensive mathematical toolkit for analysing this phenomenon and comparing its occurrences in different sectors of the economy. The basic elements of this toolkit are probably nowhere better set out than in Sorrell and Dimitropoulos' (2008) exposition of rebound effect calculus, which can serve as a basic primer on the subject.

The difficulty in this, however, is the entrenched underlying assumption that the rebound effect is entirely a price effect, and that any other reasons for lower than expected energy savings are due to technical failures and belong in a different discussion.

An effect of this entrenchment is that it sets up a framework for dealing with energy saving shortfalls that divides the world artificially into purely human actions and the actions of purely mechanical technology. Actual empirical research on energy consumption in real homes puts a big question mark alongside this human–technical division.

Fortunately the mathematics worked out by these economists do not demand a narrow focus on the price effect. Rebound effect mathematics – such as that developed by Sorrell and Dimitropoulos (2008) and that offered in the Appendix of this book – work perfectly well whatever the causes of lower than expected energy savings.

A recommendation is therefore that economists re-conceive the rebound effect in terms that fit more closely with the actual empirical situations in which it occurs, and that the notion of consumers and their home (or other appliances) as a socio-technical system would be a good place to start.

2.2 Building engineers need to take a fuller account of the socio-technical behaviour of their creations

There is a strong professional interest among many Continental building engineers to design and perfect thermal retrofits with the highest energy efficiency possible. This is part of the motivation behind the EU's Energy Performance of Buildings Directive (EPBD). The Directive requires member states to ensure that by 2021 all new buildings are so-called 'nearly zero-energy buildings' (EC, 2010). This is not demanded for existing buildings. Instead, each member state has to devise a system of energy certification for these, and conduct regular inspections of the energy efficiency of boilers and air conditioning systems. However, the energy efficiency timetable of the EPBD is used by some countries as a benchmark to guide their regulations on thermal upgrades of existing homes. In Germany's building regulations, for example, thermal standards for upgrades of existing homes have been set proportionately to, but about 30 per cent below, the steady tightening of energy efficiency requirements for new builds (Galvin, 2010, 2014c). There is naturally a strong professional interest among engineers to reach ambitious thermal standards in their retrofit designs.

However, this can be counterproductive and lead to large rebound effects if the socio-technical aspects of living in an energy efficient home are not taken fully into account. Three case studies outlined in Chapter 2 showed how both actual consumption and rebound effects became progressively larger, as the theoretical energy efficiency of apartment block retrofits increased. From a purely technical standpoint, underfloor heating is more energy efficient than standard radiators; heat pumps are more energy efficient than gas or district heating as an energy source for space heating; and finely grained thermostat and timing controls save more energy than plain dials with less nuanced controls. Often in practice, however, underfloor heating clashes with the rhythms of hour-by-hour and day-by-day heating needs of occupants; heat pumps' efficiency slumps when not ideally matched to conditions of usage; and occupants regularly fail to master sophisticated controls – the instructions and user manuals are poor; little follow up advice is offered; and many of the interfaces are overly complicated for households. Factors such as these bring practical limits to how much energy can be saved through retrofitting, and if these limits are not taken seriously then rebound effects can become very large.

Nevertheless, another side of good engineering is to strive to match technology to the real needs and abilities of people. The London office building retrofits discussed in Chapter 7 are examples of an attempt to do this from the planning stage onwards. These were not driven by an ambition to produce the theoretically highest energy efficiency in the buildings. Rather, they began and ended with discussions between occupants, building operators, owners and engineers, to work toward optimising the nuances of organisation and function between all the energy-related elements of the building and its human occupants. The key areas of concern that arise in this and a number of studies of office building energy saving are: taking into account the thermal advantages of the substance of the existing building; engaging occupants in discussion and cooperation from the planning stage forward; having a well thought-out and effective day-by-day building management system; and integrating an effective, finely grained energy monitoring system into this management.

Energy consumption in UK Government buildings may also benefit from this type of approach when the 'Government Soft Landings' mandate comes into force in 2016 (GSL, 2014). This aims to reduce performance gaps in government projects by setting and tracking targets in three aspects of sustainability: environmental, economic and social.

Life in office buildings is different from that in dwellings, but the potential exists for learning to transfer between these two domains. The main point in common is that both can be framed as socio-technical systems rather than purely material-technical edifices occupied by purely social beings.

In a recent research project in the UK in which this author was involved, it was seen how a retrofit's success can depend on the richness of interactions between the supervising architect, the owner-occupiers, and the building professionals who make the technical changes to the dwelling (Sunikka-Blank and Galvin, 2014a). This project investigated thermal retrofits of homes which have aesthetic and heritage value, where owner-occupiers were constrained between saving energy, preserving the heritage or nostalgic value of the building, and keeping within budget. These constraints forced homeowners to have to think carefully about how they wished to live in their dwellings, what aesthetic features they most wanted to preserve, and what level of thermal comfort was acceptable. These types of constraints are tailor-made for a socio-technical approach to retrofitting.

On the other hand, the passive house standard can be seen as a success story, even though a passive house is a fixed arrangement of technical features which is effectively non-negotiable with occupants. Retrofitting existing homes to passive house standard can be technically very difficult and prohibitively expensive, but new passive houses are relevant to this discussion. These dwellings' 'rebound effects' can be measured as EPGs, as discussed in Chapter 5. Even though these EPGs seem to average around 45 per cent, most passive houses still consume considerably less primary heating energy than conventional houses of the same layout. Life in a passive house has not been investigated from a socio-technical point of view, but it is interesting that heating technology in these homes is socio-technically very simple. There is no 'heating system' for occupiers to mismanage or over-use. Issues of solar incursion have been incorporated into the design. There is no negotiation on thickness of insulation, as this is standardised. Not everyone feels comfortable in a passive house, because of the limited options for control and the continuous air movement due to the ventilation heat-recovery system. An interesting area for future research would be to investigate the socio-technical aspects of living in passive houses and see whether these findings could enrich current understanding of how to plan retrofits of existing houses, not necessarily to passive house standard but using a similar technical approach.

3. Types and definitions of rebound effects

3.1 Three different metrics

Three different definitions of the rebound effect were identified in this book. Discussion in this book favours the 'elasticity rebound effect'. This is mathematically robust and flexible; it enables rebound comparisons to be made coherently between all kinds of buildings and between all the different sectors of an economy; it is not affected by different scales of energy efficiency; it can be used for time series data of national housing stocks, case studies of individual retrofits, cross-sectional 'snapshot' data of housing stocks, or (to some extent) proxy situations where price elasticities are known.

A quite different definition is the EPG, which is especially useful for new builds or situations where only the post-retrofit levels of actual and calculated consumption are known. It can also provide a cross-check as to whether the elasticity rebound effect gives a fair picture of energy savings or energy consumption impact in a particular retrofit situation. A third definition, the energy savings deficit (ESD) is useful for case study retrofits where the pre-retrofit calculated consumption is not known.

A serious problem in academic literature rests on the oft-made assumption that these three metrics are one and the same thing. This leads to unnecessary hand-wringing as to why rebound effect results from different studies of the same situations are so divergent. It also leads to inaccurate summaries of the range of 'rebound effects' occurring in housing (see further discussion in Galvin, 2014a).

Two simple recommendations are that researchers on the rebound effect in domestic heating make extra effort to make their definition of the 'rebound effect' clear, and that they refrain from using results from other researchers who have not made their own definitions clear.

Although these three rebound effect metrics are quite different from one another, an interesting point came to light regarding elasticity rebound effects and EPGs with datasets of non-residential as compared to residential buildings. Elasticity rebound effects for non-residential buildings were significantly lower than for residential, but EPGs were significantly higher. Higher EPGs means the ratio of actual to calculated consumption is high, while lower elasticity rebound effects means this ratio does not get much higher when energy efficiency increases. Therefore, a low elasticity rebound effect in itself does not necessarily imply that deep energy saving goals are being achieved.

This point can also be made in relation to fuel poverty. Fuel-poor homes have high elasticity rebound effects when retrofitted, but their pre- and post-retrofit EPGs are low (i.e. deeply negative). Hence these dwellings' high rebound effects do not necessarily imply shifts toward high absolute levels of energy consumption. It is important to take both the elasticity rebound effect and the EPG into account in rebound effect studies.

3.2 The prebound effect, the EPG and the elasticity rebound effect

The prebound effect offers a neat way of bringing the EPG and the elasticity rebound effect together in the one mathematical expression. The mathematical proof of this is shown in the Appendix, Section 10. The usefulness of the prebound effect formula in this respect was shown in Chapters 3 and 4. For example, in the large dataset of French heating consumption figures explored there, it was seen that the prebound effect is represented by the formula:

$$P = 1 - 8.2058C^{-0.4987} \tag{8.1}$$

The EPG for this housing stock is the negative of this expression, while the elasticity rebound effect is the negative of its exponent.

A direct application of this set of relationships is the relative weighting of the EPG and the elasticity rebound effect in evaluating the success or otherwise of an energy efficiency upgrade. It can also frame a brief social commentary on different sorts of upgrade cases, for example:

- Fuel-poor homes have low EPGs but high elasticity rebound effects. Low EPGs indicate they have been under-consuming their fair share of society's energy resources. High elasticity rebound effects indicate they are now catching up with the rest of the developed world and enjoying a higher level of energy services than before.
- Office buildings (according to the limited data analysed in this book) tend to act in the opposite way to fuel-poor homes: they have high EPGs and low elasticity rebound effects. Their high EPGs indicate they are consuming more than their fair share of society's energy resources, but low elasticity rebound effects indicate they do not get much worse, in this respect, after an energy efficiency upgrade.
- Average homes in countries such as Germany and France, where copious data is available, tend to have medium sized EPGs and medium sized elasticity rebound effects. In relative terms they are consuming a bit more than their fair share of society's energy resources, and probably consume a bit less than their fair share after a major thermal retrofit.
- Low energy conventional (non-passive) houses generally have very high EPGs, and appear to have high elasticity rebound effects if attempts are made to improve their energy efficiency further. The higher the theoretical energy efficiency of a conventional home, the more pronounced these effects tend to be. This tends to highlight the practical limits of effective energy efficiency increases in domestic heating. Policymakers need to

understand this point clearly. Attempts to achieve an 80 per cent savings goal will be severely limited at the high performance end. The first 30 per cent of savings are often relatively problem-free, and the costs of the upgrade measures taken to achieve them may even pay back through fuel savings. After that, each 10 per cent of savings gets progressively more difficult technically, and more expensive per kWh saved. It is the last 10–20 per cent where problems become intractable.

- Passive houses have moderate to high EPGs, but this is less of a concern as these are high percentages of small absolute numbers. The elasticity rebound effect is irrelevant when passive houses are considered as a separate group, as all these have the same calculated consumption and there are no attempts to increase their energy efficiency further.

The expression for the prebound effect, therefore, contains a rich set of information which can be useful for discussing the social and energy implications of energy efficiency issues with different types of houses or buildings.

4. Reducing the rebound effect

4.1 Is the rebound effect high enough to be a problem?

To this author's knowledge, no academic studies have ever been produced on reducing the rebound effect. Almost all existent studies either estimate how big it is, compare it between sectors, or explore its implications for energy saving and climate change mitigation. How much effort should be put into attempting to reduce the rebound effect? Is it inevitable? Is there a tolerable level of rebound effect that can simply be seen as a fact of life?

The rebound effect levels for the French and German building stocks, estimated in Chapter 4 by the cross-sectional method, were just under 50 per cent and 36 per cent respectively. A positive side to this is that at least 50 per cent of the energy efficiency improvements in the French housing stock and 64 per cent in Germany are being effectively utilised to bring energy consumption down. Another positive side is that a good amount of the 50 per cent and 36 per cent rebound effects is being utilised to improve the quality of life in homes, making cold, damp, draughty houses healthier to live in. What levels of rebound effect are tolerable and acceptable?

Some studies track the rebound effect through national or global economies, and conclude that its magnitude, of around 30 per cent, is small enough to live with (Barker *et al.*, 2007; Barker *et al.*, 2009). In their study of the macroeconomic rebound effect in the UK economy, Barker *et al.* (2007) conclude:

> Hence, the results of this study contradict the Khazzoom–Brookes postulate and support the argument that improvements in energy efficiency by producers and consumers, stimulated by government policy measures, will lead to significant reductions in energy demand and hence in greenhouse gas emissions.
>
> (p. 4945)

The Khazzoom–Brookes postulate, discussed in Chapter 1, was that energy efficiency increases lead to an increase in energy consumption, a phenomenon

Khazzoom (1980) labelled 'backfire'. Later research showed that energy efficiency improvements which are specifically designed to reduce energy consumption in the housing sector generally do succeed in doing this, though only about half to two-thirds of the improvements actually lead to reductions.

Backfire does seem to occur in some cases, however. Chapter 4 gave a time series analysis of housing sector rebound effects in the 28 EU countries plus Norway, which showed backfire happening in 5 of the 29 countries: Latvia, Denmark, Finland, Malta and Lithuania. Rebound effects were significantly above 50 per cent for four more countries: the Czech Republic, Poland, Italy and Estonia. The possible inaccuracies in the data used in this analysis were acknowledged, and also the fact that these figures include electrical appliance consumption along with home heating, but it is distinctly possible that rebound effects in some European housing sectors are very high. For former Communist countries much of this could be a catch-up response to previous fuel poverty, but for developed countries such as Denmark, Finland and Italy it must be a cause for concern.

There seem to be three main issues in the question of reducing rebound effects in domestic heating. One is the positive value of the rebound effect in reducing fuel poverty, discussed in Chapter 6 and noted above. A second is the question of how such reductions would be achieved. A third issue is whether rebound effects in domestic heating should be treated differently from those in other sectors of the economy such as transport and industry.

4.2 How can the rebound effect be reduced?

The discussion in Section 2.2 above touched on the question of how rebound effects can be reduced. It was suggested that a key to reducing rebound effects is to conceive of the dwelling and its occupants in a holistic way, as a socio-technical system. Energy efficiency upgrades can then be planned and designed for optimal fit with how the home's human and technical elements intermesh together. The aim should not be to produce the most theoretically energy efficient building, but to achieve realistic energy saving goals through optimised upgrade measures with the support of the people who actually occupy the building.

In this regard some related practical points have emerged throughout the book.

High professional standards are needed for thermal retrofit tasks such as applying wall insulation, closing off outdoor air incursion, making the most of sunshine, matching the heating system to other technical components and to occupants' needs, and designing control interfaces that are user-friendly and appropriate to realistic expectations for the heating system. Skills shortages are a major problem which governments have to address in conjunction with educational institutions.

A further point is the possibility of training or education for occupants in the use of their retrofitted home. This issue emerged in discussion of office buildings, where occupants are in a hierarchical structure and can be required to engage in learning how to get the best from the building. Households are very different from workplaces, but the occasion of a thermal upgrade can provide an opportunity to engage occupants in energy saving strategies. There are examples of citizens' initiative groups providing such services (Galvin and

Sunikka-Blank, 2014a). In the UK, the consultation process for the Green Deal could be expanded to include this. The process already includes questions being asked about household heating practices, though only to make a more accurate estimate of the likely savings from technical upgrade measures. With appropriate training and skills, Green Deal advisors could enter into a fuller consultation process with occupants which takes the full gamut of social, technical and socio-technical savings potential into account.

An effective strategy could also be to target over-consumers specifically (those showing high rebound effects after a retrofit or high EPGs in a new build), for special help in operating their retrofitted or new home economically. In the discussion of passive houses in Chapter 5 it was noted that a small proportion of households in passive houses consume an over-proportionate quantity of heating fuel, while most consume within a few percentage points of the calculated consumption or below this. The same phenomenon was observed in apartment blocks recently retrofitted to a range of thermal standards (Galvin, 2013). The mathematics show that if the highest-consuming 10–20 per cent of households reduce their consumption to the level of a dwelling's calculated consumption, this can reduce average consumption to around that level, and thereby reduce rebound effects. A government-led intervention of targeting over-consumers might be more effective than current initiatives which target households on the basis of socio-economic factors regardless of their consumption levels.

Nevertheless, even if all such avenues are exploited, the *technical* features of old homes are likely to be a continued stumbling block to achieving the ambitious level of reductions required, long term, in EU policy. No matter how good the retrofit handwork and how skilled and disciplined the occupants are at operating their heating systems economically, it can be prohibitively expensive – if it is indeed possible – to bring certain sections of the housing stock up to genuine high efficiency simply because of their geometry, material fabric and orientation to the sun.

Policymakers, therefore, need to adjust their level of energy saving ambition to what is practically feasible in retrofits of the existing building stock. Continual tightening of thermal standards for building upgrades seems to lead to high rebound effects. These are due to a combination of technical failures, mismatches between technology and occupants, and excessive comfort-taking. It is very interesting that the German government has had to shelve its long-term plans for further, ongoing tightening of thermal standards for upgrades (Galvin, 2014c). It seems that the high ambition level that drove this policy was somewhat ideologically driven rather than pragmatically grounded. Demanding more energy saving was not leading to more energy saving, and the extra costs of extreme upgrade measures led to a popular backlash against the policy.

4.3 Are rebound effects in homes as serious as those in other sectors?
In Chapter 4, Section 3.3 it was shown how rebound effects hinder efforts to reduce energy consumption in housing. The physical nature of the building stock also works against achieving deep reductions. Taking Germany as an example, the average calculated heating energy consumption rating is around 210kWh/m^{2}a (Galvin and Sunikka-Blank, 2014b). The building regulations

currently require average calculated consumption to be reduced to around 100kWh/m^{2}a for comprehensive retrofits (Galvin, 2014c). This represents an approximate doubling of energy efficiency (an increase of just over 100 per cent). This seems to be the economically feasible limit, as all initiatives to increase this efficiency target further have been shown to be impracticable (Hauser et al., 2012). If the entire building stock were upgraded by this amount on average, this would reduce consumption by just over 50 per cent *if there were no rebound effects*. With the current average rebound effect of 36 per cent, the reduction would only be 38 per cent.[1]

If the rebound effect could be reduced to half its current level, i.e. to 18 per cent, this would improve matters slightly, and give an average, nationwide reduction of 45 per cent. If 30 per cent of heating fuel were replaced by renewable energy sources, this would provide a reduction in CO_2 emissions of 62 per cent.[2] It is very difficult to see how emissions could be brought below this level. To achieve 80 per cent emissions reductions with 30 per cent renewables substitution *and zero rebound effect* would require an average energy efficiency increase of 250 per cent, i.e. the stock would need to be 350 per cent as energy efficient as at present. This would represent an average calculated consumption rating of 60kWh/m^{2}a, which is 10kWh/m^{2}a better (i.e. lower) than the current new build requirement. Again it must be noted that attempts to tighten the new build requirement have stalled, along with those for retrofits, and for the same reasons.

It may seem pessimistic to bring these figures to the fore in a book which explores how energy consumption can be reduced. However, it is very important to take a scientific approach rather than be guided by ideology or optimistic but unrealistic assertions.

Reducing the calculated average energy consumption of the housing stock to around 100kWh/m^{2}a might be possible with a highly focused and very expensive long-term effort. If rebound effects could be reduced to zero this would achieve 50 per cent reductions in energy consumption. But since fuel poverty is widespread (at least in the UK and most likely also in other western European countries) and technology is never perfect, let alone socio-technological interfaces, some rebound effects will be inevitable. Is it realistic, then, for policy to aim for reductions deeper than 50 per cent and rebound effects significantly lower than 20 per cent in the housing sector?

It is informative to compare the housing and transport sectors in this respect. Barker et al. (2007) estimate direct rebound effects from all forms of energy consumption in UK households to be 23 per cent, compared to 28 per cent for road transport (this compares with the estimate of 19.6 per cent in Chapter 4 using the time series method). The residential sector is responsible for approximately 17 per cent of UK CO_2-equivalent emissions, and the transport sector 26 per cent (DECC, 2014b). Emissions from the residential sector fell by 14.4 per cent in the years 1990–2012, or 0.61 per cent per year, while transport emissions fell by 28.3 per cent in this period, or 1.14 per cent per year.

In view of the above calculations on energy and CO_2 reduction goals it could be argued that it is justified to seek to increase energy efficiency in transport radically and reduce its rebound effects sharply, but less so in the housing sector. Today's vehicles are far more fuel-efficient than those of past decades, but the technical advances in fuel efficiency have been largely offset by the increasing horsepower and weight of newer vehicles (Ajanovic *et al.*, 2012). In the US, the journeying efficiency[3] of vehicles increased by 80 per cent in the eight years 1975–1983 (Pinto, 2009). In the 20 years 1990–2010, engine horsepower of new cars in Europe increased by around 55 per cent, thus steadily offsetting the potential gains of technical efficiency increases. The technology is available to make cars that use a fraction of current fuel consumption. These would be smaller, lighter, slower, and less densely packed with electronic extras, than today's models. Unlike energy efficient homes, such vehicles would also be cheaper than the *status quo*.[4] Since cars only last about a decade, mandating changes such as this would increase the vehicle fleet's average efficiency very rapidly, whereas most of the houses standing today will still be in use in 50 years' time and beyond.

The difficulty of reducing emissions and rebound effects in the housing stock is largely due to physical, material factors: building substance, orientation to the sun, boiler limitations, etc. High emissions and consumption in the transport sector are not due to physical, material factors but to society's support for this level of power and luxury. Transport emissions are falling faster than housing stock emissions, but that is easy, as the improvements in the technical efficiency of car engines have far outstripped the overall physical design and performance of vehicles.

Current EU policies on climate emissions reduction seek an 80 per cent reduction across all sectors of the economy by 2050. It might be argued, however, that some sectors should take more of the brunt of this than others. In view of the difficulties explored in this book, it seems technically unrealistic and socially unjust to demand 80 per cent emission reductions in the residential sector, whereas there might be strong social arguments for reductions of more than 80 per cent in transport.

Popular discourse that 'the biggest potential for emissions reduction is in the housing sector' needs to be questioned. It is certainly not correct on technical or empirical grounds. Technically, such reductions are far easier in the transport sector, and a huge legacy of empirical work shows how hard it is to achieve large reductions in energy consumption in the housing sector.

There is much potential for reducing rebound effects in domestic heating, and this book has argued that this can be better realised through a socio-technical approach to planning, design, execution and follow up of retrofits. Nevertheless it may be that a certain level of rebound effect in the housing sector needs to be tolerated and seen as normal, and the limitations of technical improvements to the existing housing stock need to be soberly assessed. It need not follow, however, that this is the case for all sectors of the economy. It may be that society needs warm houses far more than it needs fast cars.

5. Conclusion

This book's broad survey of rebound effect issues in domestic heating has led to insights and recommendations on a number of levels. Among academics, economists could adopt a broader view of what constitutes a *bona fide* rebound effect, while empirical researchers and their reviewers need to make it very clear which definition of the rebound effect they are reporting. The notion of the prebound effect has also proven useful in this discussion, as it can be used as a link between the elasticity rebound effect and the EPG. It is hoped that the mathematical Appendix will be a useful resource for academics dealing with the rebound effect and for others who wish to make robust calculations of the rebound effect in particular situations, or who are interested in its theoretical underpinnings.

For stakeholders who are planning energy efficiency upgrades in homes, the notion of a socio-technical system has been suggested as a useful framework. This brings together the human and technical aspects of home heating with all their intricate interweaving, so that upgrades can be planned and followed up in ways that closely reflect the full range of happenings and influences in relation to home heating. It would also be constructive for economists to explore whether the rebound effect might be more effectively theorised if it were couched in socio-technical terms rather than merely as a price effect.

An interesting question is whether the rebound effect in domestic heating could be significantly reduced, and how much effort should be put into this. Policy incentives, drivers, programmes and encouragement would need a radical re-think. At present, the only limitation to consumption of energy services is price – a blunt instrument and not a strong deterrent. On the other hand, a policy based on reducing consumption (e.g. personal carbon allowances or a form of carbon trading) may not yet be politically acceptable. Rationing may only work for a short period of time and in extreme situations. A more positive approach would be for policy to find a way to positively harness social practices and values. The book has made a number of observations as to how rebound effects might be mitigated through a more holistic approach to planning and post-retrofit follow up, which takes fuller account of human and human–technical interface issues, as well as giving weight to the advantages a building may already have due to its substance or position.

When all these factors have been taken into account, however, rebound effects still seem to be inevitable. This is partly due to the real need to increase energy services for fuel-poor households, partly because technical perfection does not seem possible with radically steep energy efficiency upgrades on old buildings, and partly because socio-technical configurations turn out to be different in a retrofitted building from how they were before the upgrade.

Finally, society needs to ask what its relative priorities are for climate emission reductions in various sectors of the economy. Not all sectors are equal in their savings potential and tendency for rebound.

Notes

1   This is found by first inverting equation (A18) in the Appendix, to give an average post-retrofit actual consumption of 93kWh/m²a, then using the prebound effect to calculate the average pre-retrofit actual consumption, 150kWh/m²a. Therefore average actual consumption reduces from 150 to 93kWh/m²a, or 38 per cent.
2   forty-five per cent energy reductions mean the housing stock is consuming 55 per cent of the original energy. A 30 per cent renewables substitution means it is now using 70 per cent of the fossil fuels. Hence it is now producing 0.55 x 0.7 = 0.385 times its previous CO_2 emissions. This represents a reduction of 0.615, rounded to 62 per cent.
3   The term 'journeying efficiency' means the overall efficiency of the vehicle, taking into account the efficiencies of the engine, transmission system, tyre-road running and air resistance factors.
4   The fear of being unsafe in light, less heavily armoured vehicles may be misplaced, since automatic braking technology is now available which can reduce the accident rate greatly.

REFERENCES

Ackerly K, Brager G (2013) Window signalling systems: Control strategies and occupant behaviour. *Building Research and Information* 41: 342–360.

Ajanovic A, Schipper L, Haas R (2012) The impact of more efficient but larger new passenger cars on energy consumption in EU-15 countries. *Energy* 48: 346–355.

Alcott B (2005) Jevons' paradox. *Ecological Economics* 54: 9–21.

Alfredsson E (2000) Green consumption energy use and carbon dioxide emission. PhD Thesis. Umeå University, Sweden.

Alfredsson E (2004) "Green" consumption—no solution for climate change. *Energy* 29: 513–524.

Anderson W, White V, Finney A (2012) Coping with low incomes and cold homes. *Energy Policy* 49: 40–52.

Audenaert A, De Cleyn S, Vankerckhove B (2008) Economic analysis of passive houses and low-energy houses compared with standard houses, *Energy Policy* 36: 47–55.

Aune M (2007) Energy comes home. *Energy Policy* 35: 5457–5465.

Axon C, Bright S, Dixon T, Janda Kolokotroni M (2012) Building communities: Reducing energy use in tenanted commercial property. *Building Research and Information* 40: 461–472.

Bardi U (2009) Peak oil: The four stages of a new idea. *Energy* 34: 323–326.

Barker T, Elkins P, Foxon T (2007) The macro-economic rebound effect and the UK economy. *Energy Policy* 35 (10): 4935–4946.

Barker T, Dagoumas A, Rubin J (2009) The macroeconomic rebound effect and the world economy. *Energy Efficiency* 2: 411–427.

Berger T, Amann C, Formayer H, Korjenic A, Pospichal B, Neururer C, Smutny R (2014) Impacts of urban location and climate change upon energy demand of office buildings in Vienna, Austria. *Building and Environment* 81: 258–269.

Bergman N, Hawkes A, Brett D J L, Baker P, Barton J, Blanchard R, Brandon N P, Infield D, Jardine C, Kelly N, Leach M, Matian M, Peacock A D, Staffell I, Sudtharalingam S and Woodman B (2009) UK microgeneration, Part 1: policy and behavioural aspects. *Proceedings of the Institution of Civil Engineers, Energy* 162, 23–36.

Berkhout Peter HG, Muskens Jos C, Velthuijsen Jan W (2000). Defining the rebound effect. *Energy Policy* 28 (6/7), 425–432.

Berry L, Hirst E (1983) Evaluating utility residential energy conservation programmes: an overview of an EPRI workshop. *Energy Policy* 11: 77–81.

Blight T, Coley D (2013) Sensitivity analysis of the effect of occupant behaviour on the energy consumption of passive house dwellings. *Energy and Buildings* 66: 183–192.

Boardman B (1991) *Fuel poverty: From cold homes to affordable warmth*. London: Belhaven Press.

Boardman B (2010) *Fixing fuel poverty: challenges and solutions*. London: Earthscan.

Boardman B (2012) Fuel poverty synthesis: Lessons learnt, actions needed. *Energy Policy* 49: 143–148.

Bordass B, Cohen R, Standeven M, Leaman A (2001) Assessing building performance in use 3: energy performance of probe buildings. *Building Research and Information* 29: 114–128.

Bordass B, Cohen R, Field J (2004) Energy performance of non-domestic buildings – closing the credibility gap. *International conference on improving energy efficiency in commercial buildings.* Frankfurt, Germany, 2004.

Bouzarovski S, Petrova S, Sarlamanov R (2012) Energy poverty policies in the EU: A critical perspective. *Energy Policy* 49: 76–82.

Brookes, Leonard, 1979. A low energy strategy for the UK. *Atom* 269, 73–78 (March).

Brookes, Leonard, 1990. The greenhouse effect: The fallacies in the energy efficiency solution. *Energy Policy* 18 (2), 199–201.

Brunner K-M, Spitzer M, Christanell A (2012) Experiencing fuel poverty. Coping strategies of low-income households in Vienna/Austria. *Energy Policy* 49: 53–58.

Byrne M, Bovair S (1997) A working memory model of a common procedural error. *Cognitive Science* 21(1): 31–61.

Cayla J-M, Allibe B, Laurent M-H (2010) From practices to behaviours: estimating the impact of household behaviour on space heating energy consumption, in *Proceedings of the ACEEE Summer Study on Energy Efficiency in Buildings*, Pacific Grove, CA, US, 15–20 August 2010.

Chappells H, Shove E (2007) Debating the future of comfort: environmental sustainability, energy consumption and the indoor environment. *Building Research and Information* 33: 32–40.

Chitnis M, Sorrell S, Druckman A, Firth S K and Jackson T (2013) Turning lights into flights: Estimating direct and indirect rebound effects for UK households. *Energy Policy* 55: 234–250.

Chung P, Byrne M (2008) Cue effectiveness in mitigating post-completion errors in a routine procedural task. *International Journal of Human–Computer Studies* 66: 217–232.

Chung W, Hui Y, Miu Lam Y (2006) Benchmarking the energy efficiency of commercial buildings. *Applied Energy* 83: 1–14.

DECC (Department of Energy and Climate Change) (2008) *Evaluation of the Energy Efficiency Commitment 2002–05*, Wantage, UK: Eoin Lees Energy, 2006.

DECC (Department for Energy and Climate Change) (2014a) *Green Deal approved: assessor guidance.*

DECC (Department for Energy and Climate Change) (2014b) 2013 UK Greenhouse Gas Emissions, Provisional Figures and 2012 UK Greenhouse Gas Emissions, Final Figures by Fuel Type and End-User. *DECC Statistical Release* 27 March 2014.

DENA (Deutsche Energie-Agentur) (2012) Der DENA Gebäudereport 2012. Berlin: German Energy Agency. Online resource. Available at http://issuu.com/effizienzhaus/docs/dena-geb_udereport_2012_web

DfES (Department for Education and Skills) (2004) *Schools for the future: Transforming schools: an inspirational guide to remodelling secondary schools.* London: TSO.

Druckman A, Chitnis M, Sorrell Steve, Jackson T (2011) Missing carbon reductions? Exploring rebound and backfire effects in UK households. *Energy Policy* 9: 3572–3581.

Ebel W, Grossklos M, Knissel J, Loga T, Müller K (2003) Wohnen in Passiv- und Niedrigenergiehäusern: Ein vegleichende Analyse der Nutzfaktoren am Beispiel der 'Gartenhofsiedlung Lummerlund' in Wiesbaden-Dotzheim. Darmstadt, Germany: Institut Wohnen und Umwelt.

E-Control (2013) Household Energy Price Index for Europe: October 8th 2013. Joint publication of Energie-Control (Austria), The Hungarian Energy and Public Utility Regulatory Authority, and VaasaETT Global Energy Think Tank. Available at http://www.energy-uk.org.uk/publication/finish/6/958.html (Accessed 04 March 2013).

EC (European Commission) (2010) Directive 2010/31/EU of the European Parliament and of the Council of 19 May 2010 on the energy performance of buildings. *Official Journal of the European Union*, L/153: 13–35.

EC (European Commission) (2011) Communication from the Commission to the European Parliament, the Council, the European Economic and Social Committee and the Committee of the Regions: A roadmap for moving to a competitive low carbon economy in 2050. COM(2011) 112 final. Available at: http://eur-lex.europa.eu/legal-content/EN/TXT/?uri=CELEX:52011DC0112 (Accessed 06 August 2014).

Economidou M (2011) Europe's buildings under the microscope: A country-by-country review of the energy performance of buildings. Buildings Performance Institute Europe (BPIE). Available at http://www.europeanclimate.org/documents/LR_%20CbC_study. pdf (Accessed 06 August 2014).

ENERDATA (2012) Energy efficiency trends in buildings in the EU: Lessons from the ODYSSEE MURE project. EU Commission: Intelligent Energy Europe Program. Available at: http://www.odyssee-indicators.org/publications/PDF/Buildings-brochure-2012.pdf (Accessed 30 December 2013).

ENERDATA (2014) Energy efficiency trends for households in the EU. Available at: http://www.odyssee-mure.eu/publications/efficiency-by-sector/household/household-eu.pdf (Accessed 05 July 2014).

EnEV (Energieeinsparverordnung) (2009) EnEV 2009 - Energieeinsparverordnung für Gebäude. Available at: http://www.enev-online.org/enev_2009_volltext/index.htm (Accessed 26 September 2012).

English Housing Survey (2012) English Housing Survey Headline Report 2010–11. Department for Communities and Local Government. Available at: https://www.gov.uk/government/uploads/system/uploads/attachment_data/file/6735/2084179.pdf (Accessed 11 September 2014).

Erhorn H (2007) Bedarf – Verbrauch: Ein Reizthema ohne Ende oder die Chance für sachliche Energieberatung? Fraunhofer-Institut für Bauphysik, Stuttgart. Available at: http://www.buildup.eu/publications/1810 (Accessed 20 November 2011).

EST (Energy Savings Trust) (2013) Are you a victim of rebound? Available at: http://www.energysavingtrust.org.uk/Take-action/Reduce-your-carbon-footprint/Are-you-a-victim-of-rebound (Accessed 3 April 2013).

Fawcett T, Killip G, Janda K (2013) Building Expertise: Identifying policy gaps and new ideas in housing eco-renovation in the UK and France. *ECEE Summer Study Proceedings 2013*.

Feist W (1992) Passivhaus Darmstadt Kranichstein. *Bundesbaublatt*, 2/1992.

Freire-González J (2011) Methods to empirically estimate direct and indirect rebound effect of energy-saving technological changes in households. *Ecological Modelling* 223: 32–40.

Galvin R (2010) Thermal upgrades of existing homes in Germany: The building code, subsidies, and economic efficiency. *Energy and Buildings* 42: 834–844.

Galvin R (2011) Discourse and materiality in environmental policy: The case of German Federal policy on thermal renovation of existing homes. PhD Thesis, University of East Anglia, School of Environmental Sciences, UK, January 2011.

Galvin R (2012) German Federal policy on thermal renovation of existing homes: A policy evaluation. *Sustainable Cities and Society* 4: 58–66.

Galvin R (2013) Targeting 'behavers' rather than behaviours: A 'subject-oriented' approach for reducing space heating rebound effects in low energy dwellings. *Energy and Buildings* 67: 596–607.

Galvin R (2014a) Making the 'rebound effect' more useful for performance evaluation of thermal retrofits of existing homes: Defining the 'energy savings deficit' and the 'energy performance gap'. *Energy and Buildings*, 69: 515–524.

Galvin R (2014b) Estimating broad-brush rebound effects for household energy consumption in the EU 28 countries and Norway: Some policy implications of Odyssee data. *Energy Policy* 73: 323–332.

Galvin R (2014c) Why German homeowners are reluctant to retrofit. *Building Research and Information* 42(4): 398–408.

Galvin R (2014d) Are passive houses economically viable? A reality-based, subjectivist approach to cost-benefit analyses. *Energy and Buildings* 80: 149–157. Available at: http://dx.doi.org/10.1016/j.enbuild.2014.05.025

Galvin R (2014e) Integrating the rebound effect: accurate predictors for upgrading domestic heating. Building Research and Information, DOI: 10.1080/09613218.2014.988439.

Galvin R (2015) 'Constant' rebound effects in domestic heating: Developing a cross-sectional method. *Ecological Economics* 110: 28–35.

Galvin R, Sunikka-Blank M (2013) Economic viability in thermal retrofit policies: Learning from ten years of experience in Germany. *Energy Policy* 54: 343–351.

Galvin R, Sunikka-Blank M (2014a) The UK homeowner-retrofitter as an innovator in a socio-technical system. *Energy Policy* 74: 655–662. Available at: http://dx.doi.org/10.1016/j.enpol.2014.08.013

Galvin R, Sunikka-Blank M (2014b) Disaggregating the causes of falling consumption of domestic heating energy in Germany. *Energy Efficiency* 7: 851–864.

Geddes I, Bloomer E, Allen J, Goldblatt P (2011) *The health impacts of cold homes and fuel poverty*. London: The Marmot Review Team and Friends of the Earth.

Goins J, Moezzi M (2013) Linking occupant complaints to building performance. *Building Research and Information* 41: 361–372.

Gram-Hanssen K (2010) Residential heat comfort practices: Understanding users. *Building Research and Information* 38(2): 175–186.

Gram-Hanssen K (2013) Efficient technologies or user behaviour, which is the more important when reducing households' energy consumption? *Energy Efficiency* 6(3): 447–457. doi:10.1007/s12053-012-9184-4.

Gram-Hanssen (2014) Retrofitting owner-occupied housing: Remember the people. *Building Research and Information* 42: 393–397.

Greening L, Greene D (1998) Energy use, technical efficiency, and the rebound effect: A review of the literature, Report to the U.S. Department of Energy. Denver, US: Hagler Bailly and Co.

Greening L, Greene D, Difiglio C (2000) Energy efficiency and consumption – the rebound effect – a survey. *Energy Policy* 28: 389–401.

Grub M (1990) Energy efficiency and economic fallacies. *Energy Policy* 18: 73–75.

GSL (Government Soft Landings) (2014) Government Soft Landings: Strings attached. Available at: www.building.co.uk/government-soft-landings-strings-attached/5058049.article (Accessed 13 September 2014).

Gynther L, Mikkonen I, Smits A (2012) Evaluation of European energy behavioural change programmes, *Energy Efficiency* 5: 67–82.

Haas R, Biermayr R (2000) The rebound effect for space heating: Empirical evidence from Austria. *Energy Policy* 28: 403–410.

Hamilton I, Steadman P, Bruhns H (2011) *CarbonBuzz – energy data audit*. UCL Energy Institute, UK, July.

Hanf S (2013) Dämmwahn oder Klimarettung? Vom Sinn und Unsinn der energetischen Sanierung. ZDF Television documentary (screened August 2013). Available at: http://www.zdf.de/ZDFzoom/D%C3%A4mmwahn-oder-Klimarettung-29132014.html (Accessed 28 August 2013).

Harré R (2009) Saving critical realism. *Journal for the Theory of Social Behaviour* 39(2): 129–143.

Hauser G, Maas A, Erhorn H, de Boer J, Oschatz B, Schiller H (2012) Untersuchung zur weiteren Verschärfung der energetischen Anforderungen an Gebäude mit der EnEV 2012 – Anforderungsmethodik, *Regelwerk und Wirtschaftlichkeit: BMVBS-Online-Publikation*, Nr. 05/2012. BMVS, Berlin.

Hens H (2012). *Building physics heat, air and moisture: Fundamentals and engineering exercises with examples and measurements*. Second Edition. Berlin: Ernst and Young.

Hens H, Parijs W, Deurinck M (2010) Energy consumption for heating and rebound effects. *Energy and Buildings* 42: 105–110.

Herrero S, Ürge-Vorsatz D (2012) Trapped in the heat: A post-communist type of fuel poverty. *Energy Policy* 49: 60–68.

Hong S, Oreszczyn T, Ridley I (2006) The impact of energy efficient refurbishment on the space heating fuel consumption in English dwellings. *Energy and Buildings* 38: 1171–1181.

Hong S, Paterson G, Mumovic D, Steadman P (2014) Improved benchmarking comparability for energy consumption in schools. *Building Research and Information* 42: 47–61.

Howden-Chapman P, Viggers H, Chapman R, O'Sullivan K, Telfar Barnard L, Lloyd B (2012) Tackling cold housing and fuel poverty in New Zealand: A review of policies, research, and health impacts. *Energy Policy* 49: 134–142.

Humphreys M, Hancock M (2007) Do people like to feel 'neutral'? Exploring the variation of the desired thermal sensation on the ASHRAE scale. *Energy and Buildings* 39: 867–874.

IEE (Intelligent Energy Europe) (2009) European fuel poverty and energy efficiency: Project fact sheet. European Commission. Available at: http://ec.europa.eu/intelligent/index_en.html (Accessed 11 July 2014).

Ingle A, Moezzi M, Lutzenhiser L, Diamond R (2014) Better home energy audit modelling: Incorporating inhabitant behaviours. *Building Research and Information* 42(4): 409–421.

IWU (2014) *Teilenergiekennwerte von Nichtwohngebäuden (TEK): Querschnittsanalyse der Ergebnisse der Feldphase*. Darmstadt, Germany: Institut Wohnen und Umwelt.

Jagnow K, Wolf D (2008) *Technische Optimierung und Energieeinsparung*, OPTIMUS-Hamburg City-State.

Jakob M (2006) Marginal costs and co-benefits of energy efficiency investments: The case of the Swiss residential sector, *Energy Policy* 34: 172–187.

Janda K (2014) Building communities and social potential: Between and beyond organizations and individuals in commercial properties. *Energy Policy* 67: 48–55.

Jevons, William Stanley, 1865/1965. *The Coal Question: An Inquiry Concerning the Progress of the Nation, and the Probable Exhaustion of Our Coal-mines*, 3rd edition 1905. New York: Augustus M. Kelley.

Jones Lang LaSalle (2014) *A tale of two buildings: Are EPCs a true indicator of energy efficiency?* London: JLL.

Juan Y-K, Gao P, Wang J (2010) A hybrid decision support system for sustainable office building renovation and energy performance improvement. *Energy and Buildings* 42: 290–297.

Kah O, Feist W, Pfluger R, Schnieders J, Kaufmann B, Schulz T, Bastian Z, Vilz (2008) *Bewertung energetischer Anforderungen im Lichte steigender Energiepreise für die EnEV und die KfW-Förderung*. BBR-Online-Publikation.

Kassner R, Wilkens M, Wenzel W, Ortjohan J (2010) Online-Monitoring zur Sicherstellung energetischer Zielwerte in der Baupraxis, in Paper presented at 3. EffizienzTagung Bauen & Modernisieren, Hannover, Germany, November, 19–20, 2010.

Katunsky D, Korjenic A, Katunska J, Lopusniak M, Korjenic S, Doroudiani S (2013) Analysis of thermal energy demand and saving in industrial buildings: A case study in Slovakia. *Building and Environment* 81: 138–146.

Khazzoom J Daniel (1980) Economic implications of mandated efficiency in standards or household appliances. *Energy Journal* 1 (4), 21–40.

Khazzoom J Daniel (1987) Energy saving resulting from the adoption of more efficient appliances. *Energy Journal* 8, 85–89.

Khazzoom J Daniel (1989) Energy saving resulting from the adoption of more efficient appliances: a rejoinder. *Energy Journal* 10 (1), 157–166.

Killip G (2013) Products, practices and processes: exploring the innovation potential for low-carbon housing refurbishment among small and medium-sized enterprises (SMEs) in the UK construction industry. *Energy Policy* 62: 522–530.

Kniefel J (2010) Life-cycle carbon and cost analysis of energy efficiency measures in new commercial buildings. *Energy and Buildings* 42: 333–340.

Knissel J, Feist W (1997) Passivhaus in Gross Umstadt. *Bundesbaublatt* 5/1997.

Knissel J, Loga T (2006) Vereinfachte Ermittlung von Primärenergiekennwerten. *Bauphysik*, 28(4): 270–277.

Lam J, Hui S (1996) Sensitivity analysis of energy performance of office buildings. *Building and Environment* 31: 27–39.

Laurent M-H, Allibe B, Gavin R, Hamilton I, Oreszczyn T, Tigelchaar C (2013) Back to reality: How domestic energy efficiency policies in four European countries can be improved by using empirical data instead of normative calculation. *ECEEE 2013 Summer Study Proceedings*.

Liddell C (2012) The missed exam: Conversations with Brenda Boardman. *Energy Policy* 49: 12–18.

Liddell C, Laurence C (2010) Fuel poverty and human health: A review of recent evidence. *Energy Policy* 38: 2987–2997.

Loga T, Diefenbach N and Born R (2011) Deutsche Gebäudetypologie. Beispielhafte Maßnahmen zur Verbesserung der Energieeffizienz von typischen Wohngebäuden. Institute Wohnen und Umwelt, Darmstadt, Germany. Available at: http://www.

building-typology.eu/downloads/public/docs/brochure/DE_TABULA_
TypologyBrochure_IWU.pdf (Accessed 29 January 2012).

Lovins A (1988) Energy saving resulting from the adoption of more efficient appliances: Another view. *The Energy Journal* 9: 155.

McCrae A (2008) *Renewable energy: A user's guide.* Ramsbury, UK: Crowood.

MacKenzie D, Wajcman J (eds) (1985) *The social shaping of technology.* Buckingham and Bristol, UK: Open University Press.

MacKerron G (2012) Foreword. *Energy Policy* 49: 12–18.

Madlener M, Hauertmann M (2011) Rebound effects in German residential heating: Do ownership and income matter? *FCN Working Paper No. 2/2011*, Energy Research Centre, RWTH-Aachen University.

Mahdavi A, Doppelbauer E-M (2010) A performance comparison of passive and low-energy buildings. *Energy and Buildings* 42: 1314–1319.

Maxwell D, McAndrew L (2011) *Addressing the Rebound Effect.* European Commission DG ENV: A project under the Framework contract ENV.G.4/FRA/2008/0112.

Meerbeek B, te Kulve M, Gritti T, Aarts M, van Loenen E, Aarts E (2014) Building automation and perceived control: A field study on motorized exterior blinds in Dutch offices. *Building and Environment* 79: 66–77.

Menezes A, Cripps A, Bouchlaghem D, Buswell R (2011) Predicted vs. actual energy performance of non-domestic buildings: Using post-occupancy evaluation data to reduce the performance gap. *Applied Energy* 97: 355–364.

Milne G, Boardman B (2000) Making cold homes warmer: The effect of energy efficiency improvements in low-income homes. *Energy Policy* 28: 411–424.

Molin A, Rohdin P, Moshfegh B (2011) Investigation of energy performance of newly built low-energy buildings in Sweden. *Energy and Buildings* 43: 2822–2831.

Moncaster A (2012) Constructing sustainability: Connecting the social and the technical in a case study of school building projects. PhD Thesis, University of East Anglia, UK, School of Environmental Sciences, March.

Moore R (2012) Definitions of fuel poverty: Implications for policy. *Energy Policy* 49: 19–26.

Nelson A (2008) *Globalization and Global trends in green real estate investment.* San Francisco, CA, US: RREEF Research.

Nicol J, Humphreys M (2009) New standards for comfort and energy use in buildings. *Building Research and Information* 37: 68–73.

Odyssee (2013) Odyssee energy efficiency indicators in Europe. Online resource. Available at: http://www.odyssee-indicators.org/online-indicators/ (Accessed 15 December 2013).

Olfsten T, Meier A, Lanberts R (2004) Rating the energy performance of buildings. *The International Journal of Low Energy and Sustainable Buildings* 3: 1–18.

Oschatz B, Jagnow K, Wolff D (2014) *Leitfaden zum Abgleich: Energiebedarf – Energieverbrauch.* Berlin: Deutsche Energie-Agentur.

Peper S, Feist W (2008) *Gebäudesanierung 'Passivhaus imBestand' in Ludwigshafen/ Mundenheim Messung und Beurteilungder energetischen Sanierungserfolge.* Darmstadt, Germany: Passivhaus Institut.

Pinto A (2009) *Evolution of weight, fuel consumption and CO$_2$ of automobiles.* Universidade Técnica, Lisbon: Instituto Superior Técnico.

Pivo G (2008) Responsible property investment criteria developed using the Delphi Method. *Building Research and Information* 36: 20–36.

Pivo G (2014) Unequal access to energy efficiency in US multifamily rental housing: Opportunities to improve. *Building Research and Information* 42: 551–573.

PROBE (2014a) Probe archive held by the Usable Buildings Trust (UBT). Available at: www.usablebuildings.co.uk/Pages/UBProbePublications1.html (Accessed 13 September 2014).

PROBE (2014b) PROBE post-occupancy studies. Web resource offered by CIBSE. Available at: www.cibse.org/knowledge/probe-post-occupancy-studies (Accessed 13 September 2014).

Ridley I, Clarke A, Berec J, Altamiranod H, Lewis S, Durdev M, Farr A (2013) The monitored performance of the first new London dwelling certified to the Passive House standard. *Energy and Buildings* 63: 67–78.

Risholt B, Berker T (2013) Success for energy efficient renovation of dwellings: Learning from private homeowners. *Energy Policy* 61: 1022–1030.

Schipper L (2000) Editorial: On the rebound: The interaction of energy efficiency, energy use and economic activity. An introduction. *Energy Policy* 28: 351–353.

Schneiders J, Hermelink A (2006) CEPHEUS results: Measurements and occupants' satisfaction provide evidence for Passive Houses being an option for sustainable building. *Energy Policy* 34: 151–171.

Schröder F, Altendorf L, Greller M, Boegelein T (2011) Universelle Energiekennzahlen für Deutschland: Teil 4: Spezifischer Heizenergieverbrauch kleiner Wohnhäuser und Verbrauchshochrechnung für den Gesamtwohnungsbestand. *Bauphysik*, 33(4): 243–253.

Smith K (1986) *I'm not complaining: The housing conditions of elderly private tenants.* London: SHAC.

Sorrell S (2007) *The Rebound Effect: An assessment of the evidence for economy-wide energy savings from improved energy efficiency.* Sussex Energy Group, Technology and Policy Assessment function, UK Energy Research Centre.

Sorrell S, Dimitropoulos J (2008) The rebound effect: Microeconomic definitions, limitations and extensions. *Ecological Economics* 65: 636–649.

Stafford A (2011) Long-term monitoring and performance of ground source heat pumps. *Building Research and Information* 36: 566–573.

Stevenson F, Rijal H (2010) Developing occupancy feedback from a prototype to improve housing production. *Building Research and Information* 38: 549–563.

Summerfield A, Pathan A, Lowe R (2010) Changes in energy demand from low-energy homes. *Building Research and Information* 38: 42–49.

Summerfield A, Raslan R, Lowe R, Oreszczyn T (2011) How useful are building energy models for policy? A UK perspective. *Proceedings of Building Simulation 2011: 12th Conference of International Building Performance Simulation Association*, Sydney, 14–16 November.

Sunikka-Blank M, Galvin R (2012) Introducing the prebound effect: The gap between performance and actual energy consumption. *Building Research and Information* 40(3), 260–273.

Sunikka-Blank M, Galvin R (2014) Irrational homeowners? Thermal retrofits and architectural heritage in the UK housing stock (article under review).

Telfar-Barnard L (2010) Home truths and cool admissions: New Zealand housing traits and excess winter hospitalisation, PhD Thesis. Wellington, University of Otago.

Tetlow R, Beaman P, Elmualim A, Couling K (2014) Simple prompts reduce inadvertent energy consumption from lighting in office buildings. *Building and Environment* 81: 234–242.

Tigchelaar C (2011) Obligations in the existing housing stock: Who pays the bill? *ECEEE 2011 Summer Study Proceedings*.

Tigchelaar C, Leidelmeijer K (2013) *Energiebesparing: Ein samenspel van woning en bewoner – Analyse van de module Energie WoON 2012.* Amsterdam: ECN; RTGO.

Tuominen P, Klobut K, Tolman A, Adjei A, de Best-Waldhober M (2012) Energy savings potential in buildings and overcoming market barriers in member states of the European Union. *Energy and Buildings* 51: 48–55.

Tweed C (2013) Socio-technical issues in dwelling retrofit. *Building Research and Information* 41: 551–562.

Ürge-Vorsatz D, Herrero T (2012) Building synergies between climate change mitigation and energy poverty alleviation. *Energy Policy* 49: 83–90.

Vázquez A (2013) Real buildings. The challenge of reducing HVAC energy consumption. Master's Thesis, University of Cambridge, UK.

Vlasova L, Gram-Hanssen K (2014) Incorporating inhabitants' everyday practices into domestic retrofits. *Building Research and Information* 42: 512–524.

Walberg D, Holz A, Gniechwitz T and Schulze T (2011) Wohnungsbau in Deutschland – 2011 Modernisierung oder Bestandsersatz: Studie zum Zustand und der Zukunftsfähigkeit des deutschen 'Kleinen Wohnungsbaus'. Arbeitsgemeinschaft fu r zeitgemäßes Bauen, eV, Kiel. Available at www.arge-sh.de (Accessed 3 March 2012).

Wilson C, Crane L, Chryssochoidis G (2013) *Why do people decide to renovate their homes to improve energy efficiency?* Working Paper. Norwich, UK: Tyndall Centre for Climate Change Research.

Wiltshire R (2011) Low temperature district energy, district heating, renewables integration. 16th *'Building Services, Mechanical and Building Industry Days'* International Conference, 14–15 October 2011, Debrecen, Hungary.

Yun B, Zhang J, Fujiwara A (2013) Evaluating the direct and indirect rebound effects in household energy consumption behaviour: A case study of Beijing. *Energy Policy* (in press) Available at http://dx.doi.org/10.1016/j.enpol.2013.02.024i

APPENDIX: THE MATHEMATICS OF THE REBOUND EFFECT IN DOMESTIC HEATING

A.1. The rebound effect as an elasticity

An elasticity is defined as the proportionate change in one variable as a ratio of the proportionate change in another variable. Let these variables be U and V. The elasticity of U with respect to V is denoted as η_{VU} and given by:

$$\eta_{VU} = \frac{\dfrac{\partial U}{U}}{\dfrac{\partial V}{V}} \tag{A1}$$

These are *differentials* rather than simple fractions because changes may be non-linear and the denominator variable V may have non-linear influences on the numerator variable U.

They are *partial* differentials because the changes in U may be determined by or associated with changes in other variables as well as V.

Equation (A1) may be re-written in the more conventional form of a differential equation:

$$\eta_{VU} = \frac{\partial U}{\partial V} \cdot \frac{V}{U} \tag{A2}$$

When the energy efficiency ε of an appliance or system changes, the energy consumption E and the energy services consumption S may also change. The change in *energy* consumption associated with each infinitesimal change in energy efficiency is the **energy efficiency elasticity of energy consumption** $\eta_{\varepsilon E}$. This is given by:

$$\eta_{\varepsilon E} = \frac{\partial E}{\partial \varepsilon} \cdot \frac{\varepsilon}{E} \tag{A3}$$

The change in energy *services* consumption associated with each infinitesimal change in energy efficiency is the **energy efficiency elasticity of energy services consumption** $\eta_{\varepsilon S}$. This is also called the **rebound effect** (or in this book the 'elasticity rebound effect') and is given by:

$$\eta_{\varepsilon S} = \frac{\partial S}{\partial \varepsilon} \cdot \frac{\varepsilon}{S} \qquad \text{(A4)}$$

A.2. Relationship between rebound effect and energy efficiency elasticity of energy consumption

The relationship between energy efficiency ε, energy consumption E and energy services S for any value of ε is given by:

$$S = \varepsilon \cdot E \qquad \text{(A5)}$$

Hence, from (A4) and (A5):

$$\eta_{\varepsilon S} = \frac{\partial (\varepsilon \cdot E)}{\partial \varepsilon} \cdot \frac{\varepsilon}{\varepsilon \cdot E}$$

using the product rule:

$$= \left(E \cdot \frac{\partial \varepsilon}{\partial \varepsilon} + \varepsilon \cdot \frac{\partial E}{\partial \varepsilon} \right) \cdot \frac{1}{E}$$

$$= 1 + \frac{\partial E}{\partial \varepsilon} \cdot \frac{\varepsilon}{E}$$

Hence:

$$\eta_{\varepsilon S} = 1 + \eta_{\varepsilon E} \qquad \text{(A6)}$$

This enables the rebound effect to be found if the level of energy services is not known: find the energy elasticity of energy consumption and add 1.0 to it.

An alternative way of achieving the same result is to begin with $\eta_{\varepsilon E}$ and run the proof in reverse, using the quotient rule, as in Sorrell and Dimitropoulos (2008). From (A3) and (A5):

$$\eta_{\varepsilon E} = \frac{\partial \dfrac{S}{\varepsilon}}{\partial \varepsilon} \cdot \frac{\varepsilon}{S/\varepsilon}$$

$$= \left(\frac{1}{\varepsilon} \frac{\partial S}{\partial \varepsilon} - \frac{S}{\varepsilon^2} \right) \cdot \left(\frac{\varepsilon^2}{S} \right)$$

$$= \frac{1}{\varepsilon} \frac{\partial S}{\partial \varepsilon} \cdot \frac{\varepsilon^2}{S} - 1$$

$$= \frac{\partial S}{\partial \varepsilon} \cdot \frac{\varepsilon}{S} - 1$$

Hence:

$$\eta_{\varepsilon E} = \eta_{\varepsilon S} - 1 \qquad \text{(A7)}$$

A.3. Finding the rebound effect using time series data of energy efficiency and energy consumption

The elasticity η of any parameter y with respect to any other parameter x is given by:

$$\eta_{xy} = \frac{\partial y}{\partial x} \cdot \frac{x}{y}$$

If x and y are time series with values changing exponentially with time t, these can be expressed as:

$$y = A_y \cdot B_y^t \text{ and } x = A_x \cdot B_x^t$$

The derivative can then be found parametrically:

$$\frac{\partial y}{\partial t} = A_y \cdot B_y^t \cdot \ln(B_y) \quad \text{and} \quad \frac{\partial x}{\partial t} = A_x \cdot B_x^t \cdot \ln(B_x)$$

Hence:

$$\frac{\partial y}{\partial x} = \frac{A_y \cdot B_y^t \cdot \ln(B_y)}{A_x \cdot B_x^t \cdot \ln(B_x)}$$

Hence:

$$\eta_{xy} = \frac{A_y \cdot B_y^t \cdot \ln(B_y)}{A_x \cdot B_x^t \cdot \ln(B_x)} \cdot \frac{A_x \cdot B_x^t}{A_y \cdot B_y^t}$$

Hence:

$$\eta_{xy} = \frac{\ln(B_y)}{\ln(B_x)} \tag{A8}$$

If x and y are energy efficiency ε and energy consumption E respectively, the energy efficiency elasticity of energy consumption is therefore given by:

$$\eta_{\varepsilon E} = \frac{\ln(B_E)}{\ln(B_\varepsilon)} \tag{A9}$$

The rebound effect can then be found using equation (A6):

$$\eta_{\varepsilon S} = 1 + \frac{\ln(B_E)}{\ln(B_\varepsilon)}$$

A.4. Finding the rebound effect using cross-sectional data of a large dataset of calculated consumption and actual consumption

Where the calculated consumption C and actual consumption E of a large number of dwellings or buildings are known, these can be plotted and a best-fit power curve regression line drawn. The regression line is given by:

$$E = K \cdot C^D \tag{A11}$$

The energy efficiency ε of a dwelling or building is the reciprocal of the calculated consumption C (see discussion in Galvin, 2014a). Hence:

$$E = K \cdot \varepsilon^{-D} \tag{A12}$$

From (A2) the energy efficiency elasticity of energy consumption is given by:

$$\eta_{\varepsilon E} = \frac{\partial E}{\partial \varepsilon} \cdot \frac{\varepsilon}{E} \tag{A13}$$

Substituting (A12) in (A13) gives:

$$
\begin{aligned}
\eta_{\varepsilon E} &= \frac{\partial \left(K \cdot \varepsilon^{-D} \right)}{\partial \varepsilon} \cdot \frac{\varepsilon}{K \cdot \varepsilon^{-D}} \\
&= \frac{-D.K \cdot \varepsilon^{-D-1} \cdot \varepsilon}{K \cdot \varepsilon^{-D}} \\
&= \frac{-D \cdot \varepsilon^{-1} \cdot \varepsilon^{-D} \cdot \varepsilon}{\varepsilon^{-D}} \\
&= -D
\end{aligned}
$$

Hence using (A6) the rebound effect is given by:

$$\eta_{\varepsilon S} = 1 - D \tag{A14}$$

Therefore, the rebound effect is found simply by subtracting the exponent from 1. The assumption in this proof is that, when energy efficiency upgrades (or downgrades) occur in a large number of specific buildings, their average actual/calculated consumption continues to lie on the regression line of the original dataset, i.e. on average they take on the actual/calculated consumption characteristics of already existing buildings which have the same calculated consumption as the new calculated consumption of the upgraded (or downgraded) buildings.

A.5. Finding the rebound effect for a case study dwelling or building after an energy efficiency upgrade

To find the rebound effect for a single dwelling or building that has undergone an increase in energy efficiency four data points need to be known:

- The pre-upgrade calculated consumption $C1$
- The post-upgrade calculated consumption $C2$

- The pre-upgrade actual consumption $E1$
- The post-upgrade calculated consumption $E2$

A schematic of typical positions of these data points is given in Figure A1.

Bearing in mind that elasticity is defined by means of a differential equation, a differential function deals with continuous changes and not with large discrete, one-off changes as in Figure A1. Therefore a continuous route between the points 'before' and 'after' in Figure A1 needs to be chosen, so that the differential equation can be used. This route is shown in Figure A1 as a straight line. However, the shape of this line may be chosen arbitrarily because there is, in fact, no situation between the two points. Galvin (2015) shows that by far the least complicated and most easily applied route is a power function, similar to that used in Section A.5 for cross-sectional data. This produces a single value for the solution of the differential equation (A2), i.e. a single value for the rebound effect.

As in equation (A11) the power function is given by:

$$E = K \cdot C^D \tag{A15}$$

The values for K and D are found using simultaneous equations based on the values of $C1$, $C2$, $E1$ and $E2$:

$$\frac{E_1}{E_2} = \frac{K \cdot C_1^D}{K \cdot C_2^D}$$

$$= \left(\frac{C_1}{C_2}\right)^D$$

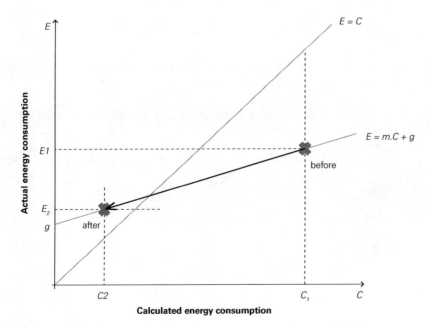

Figure A1
Schematic of typical data points of pre- and post-upgrade calculated and actual energy consumption

Hence:

$$D = \frac{\ln\left(\dfrac{E_1}{E_2}\right)}{\ln\left(\dfrac{C_1}{C_2}\right)}$$

(A16)

$$K = \frac{E_1}{C_1^D}$$

(A17)

As is shown in Section A.4, the rebound effect is given by $1 - D$. In summary:

$$\eta_{\varepsilon S} = 1 - \frac{\ln\left(\dfrac{E_1}{E_2}\right)}{\ln\left(\dfrac{C_1}{C_2}\right)}$$

(A18)

Example: A dwelling with calculated consumption $C1 = 215\text{kWh/m}^2\text{a}$ and actual consumption $E1 = 150\text{kWh/m}^2\text{a}$ is upgraded to give $C2 = 100\text{kWh/m}^2\text{a}$, and after two years running the average, weather-adjusted actual consumption $E2$ is found to be $85\text{kWh/m}^2\text{a}$. The rebound effect is therefore:

$$\eta_{\varepsilon S} = 1 - \frac{\ln\left(\dfrac{150}{85}\right)}{\ln\left(\dfrac{215}{100}\right)} = 0.2580 \; or \; 25.80\%$$

It should be noted that the energy performance gap (for definition see Section A.8) after upgrading is negative:

$$EPG = \frac{(85 - 100)}{100} = -0.15, \, or - 15\%$$

This illustrates the dangers of speaking loosely of 'the rebound effect' without being clear as to which definition is being used.

A.6. The effect on rebound effect results of changing the efficiency or calculated consumption scales

There are a number of attempts to bring the figures for calculated consumption closer to those which are typical for actual consumption (e.g. IWU, 2014). If the changes are proportional to the original calculated values, this will not affect rebound effect results. This can be proven as follows:

Energy efficiency ε is the reciprocal of calculated consumption C:

$$\varepsilon = \frac{1}{C}$$

(A19)

If the calculated consumption is adjusted by a factor g, the adjusted energy efficiency ε_1 is:

$$\varepsilon_1 = \frac{1}{(C \cdot g)} = \varepsilon \cdot g^{-1}$$

(A20)

From equation (A1), the rebound effect is defined as:

$$\eta_{\varepsilon S} = \frac{\dfrac{\partial S}{S}}{\dfrac{\partial \varepsilon}{\varepsilon}} \tag{A21}$$

With the adjusted energy efficiency this becomes:

$$\eta_{\varepsilon_i S} = \frac{\dfrac{\partial S}{S}}{\dfrac{\partial \left(\varepsilon.g^{-1}\right)}{\varepsilon.g^{-1}}} \tag{A22}$$

Since g^{-1} is a constant, this can be written as:

$$\eta_{\varepsilon_i S} = \frac{\dfrac{\partial S}{S}}{\dfrac{\left(g^{-1}\right) \cdot \partial \varepsilon}{\varepsilon.g^{-1}}}$$

$$= \frac{\dfrac{\partial S}{S}}{\dfrac{\partial \varepsilon}{\varepsilon}}$$

Hence:

$$\eta_{\varepsilon_i S} = \eta_{\varepsilon S} \tag{A23}$$

Therefore adjusting the calculated consumption by a constant scale factor does not alter the estimation of the rebound effect.

A.7. Exponential least squares regression

Time series data can be used in elasticity calculations (see A.3 above), provided there is a long-term trend over the period being considered, rather than an oscillation. The data may also show short-term, sporadic or stochastic variations during the period being considered. For the elasticity of one variable (say energy efficiency) with respect to another variable (say energy consumption) to be calculated, an average value for the annual rate of increase of each variable has to be estimated. This requires an exponential least squares regression on the time series values, of the type $y = AB^t$, where t is time, and B is the average annual rate of increase.

An easy way to find this regression is with the trend line facility on a spreadsheet such as Excel. In Excel the regression is given in the form:

$$y = A \cdot e^{Ft} \tag{A24}$$

Where e is the natural exponential (2.718281828...). Hence:

$$B = e^F \tag{A25}$$

The values of B for each of the time series are used in equations (A9) and (A10) to find the energy efficiency elasticity of energy consumption and the rebound effect.

Since these equations use the natural logarithm of the B values, a shortcut is to use the F values directly rather than the logarithm of the exponentials of the F values, since the logarithm of the exponential of a value is the value itself. Hence equation (A9) can be written as:

$$\eta_{\varepsilon E} = \frac{F_E}{F_\varepsilon} \tag{A26}$$

A further interesting point is that the F values are also the gradients of the linear regression lines of the logarithm of the time series values.

A.8. Alternative definitions of the rebound effect (a) the energy performance gap

The energy performance gap EPG is defined in this book as the excess consumption as a proportion or percentage of the calculated consumption, for the situation after a retrofit or new build. Referring to Figure A1:

$$EPG = \frac{(E_2 - C_2)}{C_2} \tag{A27}$$

A.9. Alternative definitions of the rebound effect (b) the energy savings deficit

The energy savings deficit ESD is defined in Haas and Biermayr (2000) as the shortfall in actual energy savings compared with expected energy savings, as a proportion or percentage of the expected energy savings. Expected energy savings are defined as the difference between the pre-retrofit actual consumption and the post-retrofit calculated consumption. Referring to Figure A1:

$$ESD = \frac{E_2 - C_2}{E_1 - C_2} \tag{A28}$$

Neither the ESD nor the EPG bear any relationship to the elasticity rebound effect, as the latter uses four variables, including C_1, which neither of the others utilise. 'Rebound effect' figures from studies employing different methods cannot be compared in any meaningful way.

A.10. The prebound effect

The prebound effect P was defined by Sunikka-Blank and Galvin (2012) as the shortfall in consumption compared to calculated consumption, as a proportion or percentage of the calculated consumption. It usually refers to a pre-retrofit situation or one where no energy efficiency upgrade is planned. Referring to Figure A1:

$$P = \frac{(C_1 - E_1)}{C_1} \quad \text{(A29)}$$

More generally, the prebound effect is given by:

$$P = \frac{(C - E)}{C} \quad \text{(A30)}$$

The prebound effect is, therefore, the numerical negative of the *EPG*.

For cross-sectional datasets of actual and calculated consumption (Section A.4 above), the best-fit relation between actual consumption E and calculated consumption C is usually a power curve of the type:

$$E = K \cdot C^D \quad \text{(A31)}$$

Where K and D are constants. Hence, substituting equation (A30) in equation (A31), the prebound effect is given by:

$$P = \frac{(C - K \cdot C^D)}{C} \quad \text{(A32)}$$

Hence:

$$P = 1 - K \cdot C^{D-1} \quad \text{(A33)}$$

It will be seen from equations (A12) and (A14) that the exponent $D - 1$ in equation (A33) is the numerical negative of the elasticity rebound effect. Hence the prebound effect can be used to calculate the elasticity rebound effect if it is expressed in this form.

A.11. Actual energy efficiency increases needed to meet energy consumption policy goals, given the rebound effect

Recalling equation (A9) for time series data, the energy efficiency elasticity of energy consumption is given by:

$$\eta_{\varepsilon E} = \frac{\ln(B_E)}{\ln(B_\varepsilon)} \quad \text{(A34)}$$

Hence:

$$B_\varepsilon^{\eta_{\varepsilon E}} = B_E \quad \text{(A35)}$$

Hence:

$$B_\varepsilon = \sqrt[\eta_{\varepsilon E}]{B_E} \quad \text{(A36)}$$

Substituting equation (A7) in (A36):

$$B_\varepsilon = {}^{\eta_{\varepsilon S}}\sqrt{B_E} \tag{A37}$$

Let Q be the proportionate energy efficiency increase required to meet a policy goal which aims to reduce energy consumption by N. Hence:

$$Q = B_\varepsilon - 1 \tag{A38}$$

$$N = 1 - B_E \tag{A39}$$

Substituting equations (A38) and (A39) in (A37) gives:

$$Q = {}^{\eta_{\varepsilon S}}\sqrt{1 - N} - 1 \tag{A40}$$

For example, if the rebound effect is 30 per cent and policymakers wish to reduce energy consumption by 80 per cent, energy efficiency will need to be increased by:

$$Q = {}^{0.3}\sqrt{1 - 0.8} - 1 \doteq 8.966 \ or \ 896.6\% \tag{A41}$$

Hence, the energy efficiency ε will need to be 9.966 times its original level. Since calculated energy consumption is the reciprocal of energy efficiency, this represents a required calculated consumption C of $1/9.966 = 0.103$, or 10.03% of the original level. More formally expressed:

$$C = \frac{1}{1 + Q} \tag{A42}$$

A.12. Rebound effects in more complex relations between actual and calculated consumption

Recalling equation (A4), the rebound effect is given by the differential equation:

$$\eta_{\varepsilon S} = \frac{\partial S}{\partial \varepsilon} \bullet \frac{\varepsilon}{S} \tag{A43}$$

If the rebound effect $\eta_{\varepsilon S}$ is a constant K for all values of the energy efficiency ε, the general solution to the differential equation (A43) is:

$$S = U \bullet \varepsilon^K \tag{A44}$$

This is shown by substituting (A44) in (A43):

$$\frac{\partial S}{\partial \varepsilon} \bullet \frac{\epsilon}{S} = K \bullet U \bullet \varepsilon^{K-1} \bullet \frac{\varepsilon}{U \bullet \varepsilon^K}$$

$$= K$$

Similarly, if the energy efficiency elasticity of energy consumption $\eta_{\varepsilon E}$ is a constant J, the solution to equation (A3) is:

$$E = V \bullet \varepsilon^J \tag{A45}$$

Further, since calculated consumption C is the reciprocal of energy efficiency ε:

$$E = V \bullet C^{-J} \tag{A46}$$

For this reason it is convenient if the relationship between calculated energy consumption C and actual energy consumption E is a power curve of this type.

In some cases, however, a power curve is not the relation that has the best fit with the data. For the case of the German housing stock, for example, it has been shown that a slightly better fit is obtained with the curve:

$$E = 12C^{0.499} - 29.3 \tag{A47}$$

Hence:

$$E = 12\varepsilon^{-0.499} - 29.3 \tag{A48}$$

To evaluate the rebound effect, two methods are possible from this point. The first is to find an expression for the energy efficiency of energy consumption and add 1 to it (see Section A.2), beginning by substituting (A48) in (A3):

$$
\begin{aligned}
\eta_{\varepsilon E} &= \frac{\partial E}{\partial \varepsilon} \bullet \frac{\varepsilon}{E} \\
&= \frac{\partial\left(12\varepsilon^{-0.499} - 29.3\right)}{\partial \varepsilon} \bullet \frac{\varepsilon}{12\varepsilon^{-0.499} - 29.3} \tag{A49} \\
&= \frac{-0.499 \times 12\varepsilon^{-0.499}}{12\varepsilon^{-0.499} - 29.3} \tag{A50}
\end{aligned}
$$

Hence the rebound effect is given by:

$$\eta_{\varepsilon S} = 1 + \frac{-0.499 \times 12\varepsilon^{-0.499}}{12\varepsilon^{-0.499} - 29.3} \tag{A51}$$

The second method is to express the energy services S in terms of ε and E (from equation [A5]) and begin with equation (A4). From equation (A5):

$$S = E \bullet \varepsilon \tag{A52}$$

Hence from equation (A4)

$$\eta_{\varepsilon S} = \frac{\partial S}{\partial \varepsilon} \bullet \frac{\varepsilon}{S} = \frac{\partial\left(E \bullet \varepsilon\right)}{\partial \varepsilon} \bullet \frac{\varepsilon}{E \bullet \varepsilon} \tag{A53}$$

$$= \frac{\partial\left(12\varepsilon^{-0.499+1} - 29.3\varepsilon\right)}{\partial\varepsilon} \bullet \frac{1}{12\varepsilon^{-0.499} - 29.3} \tag{A54}$$

$$= \frac{\left(1 - 0.499\right)\bullet 12\varepsilon^{-0.499} - 29.3}{12\varepsilon^{-0.499} - 29.3} \tag{A55}$$

$$= \frac{12\varepsilon^{-0.499} - 29.3 - 0.499 \times 12\varepsilon^{-0.499}}{12\varepsilon^{-0.499} - 29.3} \tag{A56}$$

$$= 1 + \frac{-0.499 \times 12\varepsilon^{-0.499}}{12\varepsilon^{-0.499} - 29.3} \tag{A57}$$

The two solutions are identical.

The rebound effect is therefore seen to vary over the range of energy efficiency. In practical terms this means that rebound effects vary depending on the pre- and post-upgrade levels of energy efficiency. To find an average rebound effect for a housing stock equation (A56/A57) has to be integrated over the stock's range of energy efficiency and the result divided by the range:

$$\overline{\eta_{\varepsilon S}} = \frac{\displaystyle\int_{\varepsilon_2}^{\varepsilon_1} \eta_{\varepsilon S}\,d\varepsilon}{\varepsilon_1 - \varepsilon_2} \tag{A58}$$

It is usually easiest to integrate a function of this type iteratively using a computer program such as Visual Basic or C++. An example is given in Galvin (2014e), using this method and the more difficult method of performing the integration of equation (A57).

References for Appendix:

Galvin R (2014a) Making the 'rebound effect' more useful for performance evaluation of thermal retrofits of existing homes: Defining the 'energy savings deficit' and the 'energy performance gap'. *Energy and Buildings*, 69: 515–524.

Galvin R (2014e) Integrating the rebound effect: accurate predictors for upgrading domestic heating. Building Research and Information, DOI: 10.1080/09613218.2014.988439.

Galvin R (2015) 'Constant' rebound effects in domestic heating: Developing a cross-sectional method. *Ecological Economics*, 110: 28–35.

Haas R, Biermayr R (2000) The rebound effect for space heating: Empirical evidence from Austria. *Energy Policy* 28: 403–410.

IWU (2014) *Teilenergiekennwerte von Nichtwohngebäuden (TEK): Querschnittsanalyse der Ergebnisse der Feldphase*. Darmstadt, Germany: Institut Wohnen und Umwelt.

Sorrell S, Dimitropoulos J (2008) The rebound effect: Microeconomic definitions, limitations and extensions. *Ecological Economics* 65: 636–649.

Sunikka-Blank M, Galvin R (2012) Introducing the prebound effect: The gap between performance and actual energy consumption. *Building Research and Information* 40(3): 260–273.

INDEX